T0262146

Robotic Manipulators

Robotic Manipulators

Edited by **Dmitry Hofland**

New York

Published by NY Research Press,
23 West, 55th Street, Suite 816,
New York, NY 10019, USA
www.nyresearchpress.com

Robotic Manipulators
Edited by Dmitry Hofland

International Standard Book Number: 978-1-63238-402-7 (Hardback)

Printed in the United States of America.

Contents

Permissions

List of Contributors

Preface

This book aims to highlight the current researches and provides a platform to further the scope of innovations in this area. This book is a product of the combined efforts of many researchers and scientists, after going through thorough studies and analysis from different parts of the world. The objective of this book is to provide the readers with the latest information of the field.

This book includes contributions of prominent researchers and engineers dealing with robotics and its related aspects. In the past few decades, interest in robotics has notably increased which has led to the advancement of theoretical research and products in this field. Robotics form an essential part of modern engineering and is associated with various other fields namely mathematics, mechanism design, computer and electric & electronics. The book extensively covers optimization modeling.

I would like to express my sincere thanks to the authors for their dedicated efforts in the completion of this book. I acknowledge the efforts of the publisher for providing constant support. Lastly, I would like to thank my family for their support in all academic endeavors.

Editor

Optimization

Nonlinear Dynamic Control and Friction Compensation of Parallel Manipulators

Weiwei Shang and Shuang Cong
University of Science and Technology of China
P.R. China

1. Introduction

Comparing with the serial ones, parallel manipulators have potential advantages in terms of high stiffness, accuracy and speed (Merlet, 2001). Especially the high accuracy and speed performances make the parallel manipulators widely applied to the following fields, like the pick-and-place operation in food, medicine, electronic industry and so on. At present, the key issues are the ways to meet the demand of high accuracy in moving process under the condition of high speed. In order to realize the high speed and accuracy motion, it's very important to design efficient control strategies for parallel manipulators.

In literatures, there are two basic control strategies for parallel manipulators (Zhang et.al., 2007): kinematic control strategies and dynamic control strategies. In the kinematic control strategies, parallel manipulators are decoupled into a group of single axis control systems, so they can be controlled by a group of individual controllers. Proportional-derivative (PD) control(Ghorbel et.al., 2000; Wu et.al., 2002), nonlinear PD (NPD) control (Ouyang et.al., 2002; Su et.al., 2004), and fuzzy control (Su et.al., 2005) all belong to this type of control strategies. These controllers do not always produce high control performance, and there is no guarantee of stability at the high speed. Unlike the kinematic control strategies, full dynamic model of parallel manipulators is taken into account in the dynamic control strategies. So the nonlinear dynamics of parallel manipulators can be compensated and better performance can be achieved with the dynamic strategies.

The traditional dynamic control strategies of parallel manipulators are the augmented PD (APD) control and the computed-torque (CT) control (Li & Wu, 2004; Cheng et.al., 2003; Paccot et.al., 2009). In the APD controller (Cheng et.al., 2003), the control law contains the tracking control term and the feed-forward compensation term. The tracking control term is realized by the PD control algorithm. The feed-forward compensation term contains the dynamic compensation calculated by the desired velocity and desired acceleration on the basis of the dynamic model. Compared with the simple PD controller, the APD controller is a tracking control method. However, the feed-forward compensation can not restrain the trajectory disturbance effectively, thus the tracking accuracy of the APD controller will be decreased. In order to solve this problem, the CT controller including the velocity feed-back is proposed based on the PD controller (Paccot et.al., 2009). The CT control method yields a controller that suppresses disturbance and tracks desired trajectories uniformly in all configurations of the manipulators. Both the APD controller and the CT controller contain two parts including the PD control term and the dynamic compensation term. For the presence of nonlinear factors such as modeling error and nonlinear friction in the dynamic models of the parallel manipulators, those traditional controllers can not achieve good control accuracy.

In order to overcome the uncertain factors in parallel manipulators, nonlinear control methods and friction compensation method are developed in this chapter. Firstly, in order to restrain the modeling error of parallel manipulators, a nonlinear PD (NPD) control algorithm is used to the APD controller, and a so-called augmented NPD (ANPD) controller is designed. Secondly, considering the feed-forward compensation term in the ANPD controller can not restrict the external disturbance, and the tracking accuracy will be affected when the disturbance exists. Thus the NPD controller is combined with the CT controller further, and a new control method named nonlinear CT (NCT) controller is developed. Thirdly, in order to compensate the nonlinear friction of parallel manipulators, a nonlinear model with two-sigmoid-function is introduced to modeling the nonlinear friction. This nonlinear friction model enables reconstruction of viscous, Coulomb, and Stribeck friction effects of parallel manipulators, and the nonlinear optimization tool is used to estimate the parameters in this model. In addition to the theoretical development, all the proposed methods in this chapter are validated on an actual parallel manipulator. The experiment results indicate that, compared with the conventional controllers, the proposed ANPD and NCT controller can get better trajectory tracking accuracy of the end-effector. Moreover, the experiment results also demonstrate that the nonlinear friction model is more accurately to compensate the friction, and is robust against the trajectory and the velocity changes.

2. Dynamic modelling

The experiment platform is a 2-DOF parallel manipulator with redundant actuation. As shown in Fig. 1, a reference frame is established in the workspace of the parallel manipulator. The unit of the frame is meter. The parallel manipulator is actuated by three servo motors located at the base A1, A2, and A3, and the end-effector is mounted at the common joint O, where the three chains meet. Coordinates of the three bases are A1 $(0, 0.25)$, A2 $(0.433, 0)$, and A3 $(0.433, 0.5)$, and all of the links have the same length $l = 0.244$ m. The definitions of the joint angles are shown in the Fig. 1, q_{a1}, q_{a2}, q_{a3} refer to the active joint angles and q_{b1}, q_{b2}, q_{b3} refer to the passive joint angles.

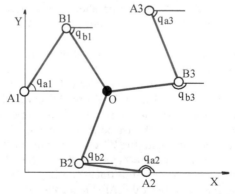

Fig. 1. Coordinates of the 2-DOF parallel manipulator with redundant actuation

Cutting the parallel manipulator at the common point O in Fig. 1, one can have an open-chain system including three independent planar 2-DOF serial manipulators, each of which contains an active joint and a passive joint. The dynamic model of the parallel manipulator equals to the model of the open-chain system plus the closed-loop constraints, thus the dynamic model of the whole parallel manipulator can be formulated by combining the dynamics of the three serial manipulators under the constraints.

As we know, the dynamic model of each planar 2-DOF serial manipulator can be formulated as (Murray et.al., 1994)

$$\mathbf{M}_i \ddot{\mathbf{q}}_i + \mathbf{C}_i \dot{\mathbf{q}}_i + \mathbf{f}_i = \boldsymbol{\tau}_i \tag{1}$$

where $\mathbf{q}_i = \begin{bmatrix} q_{ai} & q_{bi} \end{bmatrix}^T$, q_{ai} and q_{bi} are the active joint and passive joint angle, respectively; $\mathbf{M}_i$ is inertia matrix, and $\mathbf{C}_i$ is Coriolis and centrifugal force matrix, which are defined as

$$\mathbf{M}_i = \begin{bmatrix} \alpha_i & \gamma_i \cos(q_{ai} - q_{bi}) \\ \gamma_i \cos(q_{ai} - q_{bi}) & \beta_i \end{bmatrix}$$

$$\mathbf{C}_i = \begin{bmatrix} 0 & \gamma_i \sin(q_{ai} - q_{bi}) \dot{q}_{bi} \\ -\gamma_i \sin(q_{ai} - q_{bi}) \dot{q}_{ai} & 0 \end{bmatrix}$$

where $\alpha_i, \beta_i, \gamma_i, i = 1,2,3$ are the dynamic parameters which are related with the physical parameters such as mass, center of mass, and inertia. In Eq.(1), $\boldsymbol{\tau}_i = \begin{bmatrix} \tau_{ai} & \tau_{bi} \end{bmatrix}^T$ is joint torque vector, where τ_{ai} is the active joint torque, the passive joint torque $\tau_{bi} = 0$. Vector $\mathbf{f}_i = \begin{bmatrix} f_{ai} & f_{bi} \end{bmatrix}^T$ is the friction torque, where f_{ai} and f_{bi} are the active joint friction and passive joint friction, respectively. The friction parameters of the active joints and the passive joints are identified simultaneously for the parallel manipulator (Shang et.al., 2010). And from the identified results, one can find that the friction parameters of the passive joints are much smaller than those of the active joints. Thus, compared with the active joints friction f_{ai}, the passive joint friction f_{bi} is much smaller and it can be neglected (Shang et.al., 2010). Generally, the active joint friction torque f_{ai} can be formulated by using the Coulomb + viscous friction model as

$$f_{ai} = sign(\dot{q}_{ai}) f_{ci} + f_{vi} \dot{q}_{ai} \tag{2}$$

where f_{ci} represents the Coulomb friction, and f_{vi} represents the coefficient of the viscous friction.

Combining the dynamic models of three 2-DOF serial manipulators, the dynamic model of the open-chain system can be expressed as

$$\mathbf{M}\ddot{\mathbf{q}} + \mathbf{C}\dot{\mathbf{q}} + \mathbf{f} = \boldsymbol{\tau} \tag{3}$$

where the definition of the symbols is similar to those in Eq.(1), only the difference is that the symbols in Eq.(3) represent the whole open-chain system not a 2-DOF serial manipulator. Based on Eq.(3) of the open-chain system and the constraint forces due to the closed-loop constraints, the dynamic model of the parallel manipulator can be written as

$$\mathbf{M}\ddot{\mathbf{q}} + \mathbf{C}\dot{\mathbf{q}} + \mathbf{f} = \boldsymbol{\tau} + \mathbf{A}^T \boldsymbol{\lambda} \tag{4}$$

where $\mathbf{A}^T\lambda$ represents the constraint force vector, here matrix $\mathbf{A}$ is the differential of the closed-loop constrained equation and λ is a unknown multiplier representing the magnitude of the constraint forces. Fortunately, $\mathbf{A}^T\lambda$ can be eliminated, by finding the null-space of matrix $\mathbf{A}$ (Muller, 2005). With the Jacobian matrix $\mathbf{W}$, we have

$$\dot{\mathbf{q}} = \mathbf{W}\dot{\mathbf{q}}_e \tag{5}$$

where $\dot{\mathbf{q}} = \begin{bmatrix} \dot{q}_{a1} & \dot{q}_{a2} & \dot{q}_{a3} & \dot{q}_{b1} & \dot{q}_{b2} & \dot{q}_{b3} \end{bmatrix}^T$ represents the velocity vector of all the joints, $\dot{\mathbf{q}}_e = \begin{bmatrix} \dot{q}_x & \dot{q}_y \end{bmatrix}$ represents the velocity vector of the end-effector, and the Jacobian matrix $\mathbf{W}$ is defined as

$$\mathbf{W} = \begin{bmatrix} r_1\cos(q_{b1}) & r_1\sin(q_{b1}) \\ r_2\cos(q_{b2}) & r_2\sin(q_{b2}) \\ r_3\cos(q_{b3}) & r_3\sin(q_{b3}) \\ -r_1\cos(q_{a1}) & -r_1\sin(q_{a1}) \\ -r_2\cos(q_{a2}) & -r_2\sin(q_{a2}) \\ -r_3\cos(q_{a3}) & -r_3\sin(q_{a3}) \end{bmatrix}, \text{ where } r_i = \frac{1}{l\sin(q_{bi} - q_{ai})}$$

Considering the constraint equation $\mathbf{A}\dot{\mathbf{q}} = 0$, then one can have $\mathbf{A}\mathbf{W}\dot{\mathbf{q}}_e = 0$ with the Jacobian relation Eq.(5). The velocity vector $\dot{\mathbf{q}}_e$ of the end-effector contains independent generalized coordinates, so one can get $\mathbf{A}\mathbf{W} = 0$, or equivalently, $\mathbf{W}^T\mathbf{A}^T = 0$. With this result, the term of $\mathbf{A}^T\lambda$ can be eliminated, and the dynamic model Eq. (4) can be written as

$$\mathbf{W}^T\mathbf{M}\ddot{\mathbf{q}} + \mathbf{W}^T\mathbf{C}\dot{\mathbf{q}} + \mathbf{W}^T\mathbf{f} = \mathbf{W}^T\tau + \mathbf{W}^T\mathbf{A}^T\lambda = \mathbf{W}^T\tau \tag{6}$$

In order to study dynamic control and trajectory planning of the parallel manipulator both in the task space, we will further formulate the dynamic model in the task space on the basis of the dynamic model Eq. (6) of the joint space. Differentiating the Jacobian Eq. (5) yields

$$\ddot{\mathbf{q}} = \dot{\mathbf{W}}\dot{\mathbf{q}}_e + \mathbf{W}\ddot{\mathbf{q}}_e \tag{7}$$

and substituting Eqs. (5) and (7) into Eq. (6), the dynamic model in the task space can be written as

$$\mathbf{W}^T\mathbf{M}\mathbf{W}\ddot{\mathbf{q}}_e + \mathbf{W}^T(\mathbf{M}\dot{\mathbf{W}} + \mathbf{C}\mathbf{W})\dot{\mathbf{q}}_e + \mathbf{W}^T\mathbf{f} = \mathbf{W}^T\tau \tag{8}$$

If the friction torques of the passive joints is neglected, then Eq. (8) can be further simplified. Let τ_a and $\mathbf{f}_a$ be the actuator and friction torque vector of the three active joints respectively, then $\mathbf{W}^T\tau = \mathbf{S}^T\tau_a$, and $\mathbf{W}^T\mathbf{f} = \mathbf{S}^T\mathbf{f}_a$. Here, $\mathbf{S}$ is the Jacobian matrix between the velocity of the end-effector and the velocity of three active joints, and $\mathbf{S}$ is written as

$$\mathbf{S} = \begin{bmatrix} r_1\cos(q_{b1}) & r_1\sin(q_{b1}) \\ r_2\cos(q_{b2}) & r_2\sin(q_{b2}) \\ r_3\cos(q_{b3}) & r_3\sin(q_{b3}) \end{bmatrix}$$

Then, the dynamic model in the task space can be written as

$$\mathbf{W}^T \mathbf{MW} \ddot{\mathbf{q}}_e + \mathbf{W}^T (\mathbf{M}\dot{\mathbf{W}} + \mathbf{CW})\dot{\mathbf{q}}_e + \mathbf{S}^T \mathbf{f}_a = \mathbf{S}^T \boldsymbol{\tau}_a \tag{9}$$

The above Eq.(9) can be briefly expressed as

$$\mathbf{M}_e \ddot{\mathbf{q}}_e + \mathbf{C}_e \dot{\mathbf{q}}_e + \mathbf{S}^T \mathbf{f}_a = \mathbf{S}^T \boldsymbol{\tau}_a \tag{10}$$

where $\mathbf{M}_e = \mathbf{W}^T \mathbf{MW}$ is the inertial matrix in the task space, and $\mathbf{C}_e = \mathbf{W}^T (\mathbf{M}\dot{\mathbf{W}} + \mathbf{CW})$ is the Coriolis and centrifugal force matrix in the task space.

The dynamic model Eq. (10) in the task space also satisfies the similar structural properties to the dynamic model of the open-chain system and the 2-DOF serial manipulator as follows (Cheng et.al., 2003):

a.　$\mathbf{M}_e$ is symmetric and positive.

b.　$\dot{\mathbf{M}}_e - 2\mathbf{C}_e$ is skew-symmetric matrix.

3. Nonlinear dynamic control by using the NPD

There are two conventional dynamic controllers for parallel manipulators: APD controller and CT controller. The common feature of the two controllers is eliminating the tracking error by linear PD control. However, the linear PD control is not robust against the uncertain factors such as modeling error and external disturbance. To overcome this problem, the NPD control can be combined with the conventional control strategies to improve the control accuracy and disturbance rejection ability.

3.1 NPD controller

As well as we know, the linear PD controller takes the form

$$u_L(t) = k_p e(t) + k_d \dot{e}(t) \tag{11}$$

where k_p and k_d are the proportional and derivative constants respectively, and $e(t)$ is the system error.

The nonlinear PD (NPD) controller has a similar structure as the linear PD controller (11), the NPD controller may be any control structure of the form

$$u_N(t) = k_p(\cdot)e(t) + k_d(\cdot)\dot{e}(t) \tag{12}$$

where $k_p(\cdot)$ and $k_d(\cdot)$ are the time-varying proportional and derivative gains, which may depend on system state, input or other variables.

Currently, several NPD controllers have been proposed for robotic application (Xu et.al., 1995; Kelly & Ricardo, 1996; Seraji et.al., 1998). The NPD controller has superior trajectory tracking and disturbance rejection ability compared with the linear PD controllers for robot control. The NPD controller proposed by Han has a simple structure as (Han, 1994)

$$u_H(t) = k_p fun(e(t), \alpha_1, \delta_1) + k_d fun(\dot{e}(t), \alpha_2, \delta_2) \tag{13}$$

where the function fun can be defined as

$$fun(x,\alpha,\delta) = \begin{cases} |x|^\alpha \, sign(x), & |x| > \delta \\ x/\delta^{1-\alpha}, & |x| \le \delta \end{cases} \tag{14}$$

where α refers to the nonlinearity, specially the NPD will degenerate into the linear PD when $\alpha = 1$; δ refers to the threshold of the error (or error derivative), and it is at the same magnitude with the error (or error derivative). The NPD controller (13) can be rewritten as the form (12), then $k_p(\cdot)$ can be derived as

$$k_p(e) = \begin{cases} k_p |e|^{\alpha_1 - 1} & |e| > \delta_1 \\ k_p \delta_1^{\alpha_1 - 1} & |e| \le \delta_1 \end{cases} \tag{15}$$

Similarly, $k_d(\cdot)$ can be expressed as

$$k_d(\dot{e}) = \begin{cases} k_d |\dot{e}|^{\alpha_2 - 1} & |\dot{e}| > \delta_2 \\ k_d \delta_2^{\alpha_2 - 1} & |\dot{e}| \le \delta_2 \end{cases} \tag{16}$$

In (15) and (16), α_1 and α_2 can be determined in the interval [0.5, 1.0] and [1.0, 1.5], respectively. This choice makes the nonlinear gains with the following characteristics (Han, 1994): on one hand, large gain for small error and small gain for large error; on the other hand, large gain for large error rate and small gain for small error rate. Such variations of the gains result in a rapid transition of the systems with favorable damping. In addition, the NPD controller is robust against the changes of the system parameters and the nonlinear factors. Thus the NPD controller (13) is suitable to the trajectory tracking of the high-speed planar parallel manipulator.

3.2 Augmented NPD controller
The augmented NPD (ANPD) controller developed here is designed by replacing the linear PD in the APD controller with the NPD algorithm. According to the APD controller and the NPD control algorithm (13), based on the dynamic model (10), the control law of the ANPD controller can be written as (Shang et.al., 2009)

$$\tau_A = \mathbf{M}_e \ddot{\mathbf{q}}_e^d + \mathbf{C}_e \dot{\mathbf{q}}_e^d + \mathbf{S}^T \mathbf{f}_a + \mathbf{K}_p(e)e + \mathbf{K}_d(\dot{e})\dot{e} \tag{17}$$

where $\dot{\mathbf{q}}_e^d$ and $\ddot{\mathbf{q}}_e^d$ are the desired velocity and acceleration of the end-effector. The control law (17) can be divided into three terms according to different functions. The first term is the dynamics compensation defined by the desired trajectory, which can be written as

$$\tau_{A1} = \mathbf{M}_e \ddot{\mathbf{q}}_e^d + \mathbf{C}_e \dot{\mathbf{q}}_e^d \tag{18.a}$$

The second term is the friction compensation, which can be written as

$$\tau_{A2} = \mathbf{S}^T \mathbf{f}_a \tag{18.b}$$

The third term is the tracking error elimination, which can be written as

$$\tau_{A3} = \mathbf{K}_p(e)e + \mathbf{K}_d(\dot{e})\dot{e} \tag{18.c}$$

where $e = q_e^d - q_e$ is the position error of the end-effector; $K_p(e)$ and $K_d(\dot{e})$ are symmetric, positive definite matrices of time-varying gains. From (15) and (16), $K_p(e)$ and $K_d(\dot{e})$ can be expressed as

$$K_p(e) = diag\left(k_p|x_1|^{\alpha_1 - 1}, k_p|x_2|^{\alpha_1 - 1}\right) \tag{19}$$

$$K_d(\dot{e}) = diag\left(k_d|y_1|^{\alpha_2 - 1}, k_d|y_2|^{\alpha_2 - 1}\right) \tag{20}$$

where k_p and k_d are the positive constant gains. The variables x_i, y_i, $i = 1,2$ are determined by the following rules: if $|e_i| > \delta_1$, then $x_i = e_i$, else $x_i = \delta_1$; if $|\dot{e}_i| > \delta_2$, then $y_i = \dot{e}_i$, else $y_i = \delta_2$; $\alpha_1, \alpha_2, \delta_1$, and δ_2 are the designed parameters which should be tuned in practice.

In the following, we will prove the asymptotic stability of the parallel manipulator system controlled by the ANPD controller (17). Firstly, we will introduce two lemmas (Kelly and Ricardo, 1996).

Lemma 1: Let $\alpha(\cdot)$ be a class K function and $f : \Re \to \Re$ a continuous function. If $f(x) \geq \alpha(|x|)$ $\forall x \in \Re$, then $\int_0^x f(\sigma)d\sigma > 0, \forall x \neq 0 \in \Re$ and $\int_0^x f(\sigma)d\sigma \to \infty$ as $|x| \to \infty$.

Lemma 2: Consider the continuous diagonal matrix $K_p : \Re^2 \to \Re^{2 \times 2}$

$$K_p(e) = \begin{bmatrix} k_{p1}(e_1) & 0 \\ 0 & k_{p2}(e_2) \end{bmatrix}$$

Assume that there exist class K functions $\alpha_i(\cdot)$ such that

$$x k_{pi}(x) \geq \alpha_i(|x|), \ x \in \Re, i = 1,2$$

then $\int_0^e \xi^T K_p(\xi)d\xi > 0, \forall e \neq 0 \in \Re^2$, and $\int_0^e \xi^T K_p(\xi)d\xi \to \infty$ as $|e| \to \infty$.

Next, we will give brief proof for Lemma 2 (Kelly and Ricardo, 1996). Define $f(e_i) = k_{pi}(e_i)e_i$, From Lemma 1, one can get

$$\int_0^{e_i} f(\xi_i)d\xi_i > 0, \ \ \forall e_i \neq 0 \in \Re \tag{21}$$

which is equivalent to

$$\int_0^{e_i} k_{pi}(\xi_i)\xi_i d\xi_i > 0, \ \ \forall e_i \neq 0 \in \Re \tag{22}$$

Therefore, the function $\int_0^e \xi^T K_p(\xi)d\xi$ is positive definite. Also, Lemma 1 ensures that above integral is radically unbounded with respect to e, and this implies $\int_0^e \xi^T K_p(\xi)d\xi \to \infty$ as $|e| \to \infty$.

Theorem 1: If the nonlinear gains $K_p(\cdot)$ and $K_d(\cdot)$ are defined by (19) and (20) respectively, the parallel manipulator system controlled by the ANPD control law (17) is asymptotically stable.

Proof: Choose the Lyapunov function candidate as

$$V(\mathbf{e},\dot{\mathbf{e}}) = \frac{1}{2}\dot{\mathbf{e}}^T\mathbf{M}_e\dot{\mathbf{e}} + \int_0^{\mathbf{e}}\boldsymbol{\xi}^T\mathbf{K}_p(\boldsymbol{\xi})d\boldsymbol{\xi} \tag{23}$$

where

$$\int_0^{\mathbf{e}}\boldsymbol{\xi}^T\mathbf{K}_p(\boldsymbol{\xi})d\boldsymbol{\xi} = \int_0^{e_1}\xi_1 k_{p1}(\xi_1)d\xi_1 + \int_0^{e_2}\xi_2 k_{p2}(\xi_2)d\xi_2$$

Considering the structural properties (a), the inertial matrix $\mathbf{M}_e$ is symmetric and positive definite matrix, thus the first term in (23) is positive definite. In addition, the integral term can be interpreted as a potential energy induced by the position error-driven part of the controller. Next, we will proof that the second term in (23) is positive definite. Considering $k_{pi}(e_i)$ is defined as

$$k_{pi}(e_i) = \begin{cases} k_{pi}|e_i|^{\alpha_i - 1}, & |e_i| > \delta \\ k_{pi}\delta_i^{\alpha_i - 1}, & |e_i| \le \delta_i \end{cases} \tag{24}$$

Define class K functions $\alpha_i(\cdot)$ as

$$\alpha_i(|e_i|) = \begin{cases} \varepsilon_i e_i |e_i|^{\alpha_i - 1}, & |e_i| > \delta_1 \\ \varepsilon_i e_i \delta_i^{\alpha_i - 1}, & |e_i| \le \delta_1 \end{cases} \quad,\text{ and } k_{pi} > \varepsilon_i > 0 \tag{25}$$

With the Lemma 2, one can get the integral term in (23) is a radically unbounded positive definite function. Thus $V(\mathbf{e},\dot{\mathbf{e}})$ is a positive function. Differentiating $V(t)$ with respect to time yields

$$\dot{V}(\mathbf{e},\dot{\mathbf{e}}) = \dot{\mathbf{e}}^T\mathbf{M}_e\ddot{\mathbf{e}} + \frac{1}{2}\dot{\mathbf{e}}^T\dot{\mathbf{M}}_e\dot{\mathbf{e}} + \mathbf{e}^T\mathbf{K}_p(\mathbf{e})\dot{\mathbf{e}} \tag{26}$$

Combine the control law (17) and the dynamic model (10), the closed-loop system equation can be written as

$$\mathbf{M}_e\ddot{\mathbf{e}} + \mathbf{C}_e\dot{\mathbf{e}} + \mathbf{K}_p(\cdot)\mathbf{e} + \mathbf{K}_d(\cdot)\dot{\mathbf{e}} = 0 \tag{27}$$

Multiplying both sides of the above equation by $\dot{\mathbf{e}}^T$, and then substituting the resulting equation into (26) yields

$$\dot{V} = -\dot{\mathbf{e}}^T\mathbf{K}_d(\cdot)\dot{\mathbf{e}} + \frac{1}{2}\dot{\mathbf{e}}^T(\dot{\mathbf{M}}_e - 2\mathbf{C}_e)\dot{\mathbf{e}} \tag{28}$$

Considering the structural properties (b), then one can have $\dot{\mathbf{e}}^T(\dot{\mathbf{M}}_e - 2\mathbf{C}_e)\dot{\mathbf{e}} = 0$ and

$$\dot{V} = -\dot{\mathbf{e}}^T\mathbf{K}_d(\cdot)\dot{\mathbf{e}} \tag{29}$$

As $\mathbf{K}_d(\cdot)$ is a symmetric, positive definite matrix, then $\dot{V}$ is a semi-negative definite matrix, thus the parallel manipulator system is stable.

Now since $V(t) \geq 0$ and $\dot{V}(t) \leq 0$, $V(t)$ is bounded and decreasing, thus $V(t)$ converges to a limit. From the definition of $V(t)$, it implies that both $\mathbf{e}$ and $\dot{\mathbf{e}}$ are bounded. Since $\mathbf{M}_e$ is uniform positive definite, then $\mathbf{M}_e^{-1}$ exists and bounded, thus the closed-loop system equation (27) can be written as

$$\ddot{\mathbf{e}} = -\mathbf{M}_e^{-1}\left(\mathbf{C}_e\dot{\mathbf{e}} + \mathbf{K}_p(\cdot)\mathbf{e} + \mathbf{K}_d(\cdot)\dot{\mathbf{e}}\right) \tag{30}$$

So $\ddot{\mathbf{e}}$ is also bounded and $\dot{V}(t)$ is bounded. Thus, $\dot{V}(t)$ is uniformly continuous. With the Barbalat Lemma (Slotine & Li, 1991), one knows $\dot{\mathbf{e}} \to 0$ as $t \to \infty$, and this implies $\mathbf{e} \to 0$ as $t \to \infty$.

One can note that $\boldsymbol{\tau}_A$ in the control law (17) is the actuator torque of the task space, but in fact, we need the actuator torque $\boldsymbol{\tau}_a$ of the active joints. In practice, a solution that has a minimum weighted Euclidian norm is selected as the actual control input. The actual control input vector of the active joints can be written as

$$\boldsymbol{\tau}_a = (\mathbf{S}^T)^+ \left(\mathbf{M}_e\ddot{\mathbf{q}}_e^d + \mathbf{C}_e\dot{\mathbf{q}}_e^d + \mathbf{K}_p(\mathbf{e})\mathbf{e} + \mathbf{K}_d(\dot{\mathbf{e}})\dot{\mathbf{e}}\right) + \mathbf{f}_a \tag{31}$$

where $(\mathbf{S}^T)^+ = \mathbf{S}(\mathbf{S}^T\mathbf{S})^{-1}$ is the pseudo-inverse of $\mathbf{S}^T$, satisfying $\mathbf{S}^T(\mathbf{S}^T)^+ = \mathbf{I}$. For the parallel manipulator with redundant actuation, the singularity is eliminated in the effective workspace (Shang et.al., 2010). Thus, the pseudo-inverse matrix $(\mathbf{S}^T)^+$ will not be close to the singularity for this parallel manipulator with redundant actuation.

3.3 Nonlinear computed torque control

An obvious drawback of the traditional CT controllers is the elimination of the tracking error by linear PD algorithm. However, the linear PD algorithm is not robust against the uncertain factors such as modeling error and nonlinear friction. To overcome this problem, the NPD algorithm can be combined with the conventional control strategies to improve the control accuracy. The NCT controller developed in this chapter is designed by replacing the linear PD in the CT controller with the NPD algorithm.

According to the NPD algorithm (13), based on the dynamic model (10), the control law of the NCT controller can be written as (Shang & Cong, 2009)

$$\boldsymbol{\tau}_N = \mathbf{M}_e\ddot{\mathbf{q}}_e^d + \mathbf{C}_e\dot{\mathbf{q}}_e + \mathbf{S}^T\mathbf{f}_a + \mathbf{M}_e\left(\mathbf{K}_p(\mathbf{e})\mathbf{e} + \mathbf{K}_d(\dot{\mathbf{e}})\dot{\mathbf{e}}\right) \tag{32}$$

The control law (32) can be divided into three terms according to the different functions. The first term is the dynamics compensation defined by the desired acceleration and the actual velocity of the end-effector, which can be written as

$$\boldsymbol{\tau}_{N1} = \mathbf{M}_e\ddot{\mathbf{q}}_e^d + \mathbf{C}_e\dot{\mathbf{q}}_e \tag{33.a}$$

The second term is the friction compensation, which can be written as

$$\boldsymbol{\tau}_{N2} = \mathbf{S}^T\mathbf{f}_a \tag{33.b}$$

The third term is the tracking error elimination, which can be written as

$$\boldsymbol{\tau}_{N3} = \mathbf{M}_e\left(\mathbf{K}_p(\mathbf{e})\mathbf{e} + \mathbf{K}_d(\dot{\mathbf{e}})\dot{\mathbf{e}}\right) \tag{33.c}$$

where $\mathbf{K}_p(\mathbf{e})$ and $\mathbf{K}_d(\dot{\mathbf{e}})$ are symmetric, positive definite matrices of time-varying gains. From (15) and (16), $\mathbf{K}_p(\mathbf{e})$ and $\mathbf{K}_d(\dot{\mathbf{e}})$ can be expressed as

$$\mathbf{K}_p(\mathbf{e}) = diag\left(k_{p1}\left|x_1\right|^{\alpha_1-1}, k_{p2}\left|x_2\right|^{\alpha_1-1}\right) \tag{34}$$

$$\mathbf{K}_d(\dot{\mathbf{e}}) = diag\left(k_{d1}\left|y_1\right|^{\alpha_2-1}, k_{d2}\left|y_2\right|^{\alpha_2-1}\right) \tag{35}$$

where k_{pi}, k_{di}, $i=1,2$ are positive constant gains. The variables x_i, y_i are determined by the following rules: if $\left|e_i\right| > \delta_1$, then $x_i = e_i$, else $x_i = \delta_1$; if $\left|\dot{e}_i\right| > \delta_2$, then $y_i = \dot{e}_i$, else $y_i = \delta_2$. $\alpha_1, \alpha_2, \delta_1$, and δ_2 are the designed parameters which should be tuned in practice.

In the following, the asymptotic stability of the parallel manipulator system controlled by the NCT controller (32) will be proven.

Theorem 2: If the nonlinear gains $\mathbf{K}_p(\cdot)$ and $\mathbf{K}_d(\cdot)$ are defined by (34) and (35) respectively, the parallel manipulator system controlled by the NCT controller (32) is asymptotically stable.

Proof: Choose the Lyapunov function candidate as

$$V(\mathbf{e}, \dot{\mathbf{e}}) = \frac{1}{2}\dot{\mathbf{e}}^T\dot{\mathbf{e}} + \int_0^{\mathbf{e}}\left|\xi\right|^T K_p(\xi)d\xi \tag{36}$$

where $\int_0^{\mathbf{e}}\left|\xi\right|^T \mathbf{K}_p(\xi)d\xi = \int_0^{e_1}\left|\xi_1\right|k_{p1}(\xi_1)d\xi_1 + \int_0^{e_2}\left|\xi_2\right|k_{p2}(\xi_2)d\xi_2$. Obviously, the first term in (36) is positive definite. In addition, the integral term can be interpreted as the potential energy induced by the position error-driven part of the controller. Next, one can prove that the second term in (36) is positive definite. Considering $k_{pi}(e_i)$ is defined as

$$k_{pi}(e_i) = \begin{cases} k_{pi}\left|e_i\right|^{\alpha_i-1}, \left|e_i\right| > \delta_1 \\ k_{pi}\delta_i^{\alpha_i-1}, \left|e_i\right| \le \delta_1 \end{cases} \tag{37}$$

and define class K functions $\alpha_i(\cdot)$ as

$$\alpha_i(\left|e_i\right|) = \begin{cases} \varepsilon_i\left|e_i\right|^{\alpha_i}, & \left|e_i\right| > \delta_1 \\ \varepsilon_i\left|e_i\right|\delta_i^{\alpha_i-1}, & \left|e_i\right| \le \delta_1 \end{cases}, \text{ and } k_{pi} > \varepsilon_i > 0 \tag{38}$$

From (37) and (38), one knows $\left|e_i\right|k_{pi}(e_i) \ge \alpha_i(\left|e_i\right|)$. With the Lemma 2, one can get $\int_0^{e_i}\left|\xi_i\right|^T k_{p_i}(\xi_i)d\xi_i > 0$, and $\int_0^{\mathbf{e}}\left|\xi\right|^T \mathbf{K}_p(\xi)d\xi \to \infty$ as $\left|\mathbf{e}\right| \to \infty$. So one can get the integral term in (36) is a radically unbounded positive definite function. Thus $V(\mathbf{e}, \dot{\mathbf{e}})$ is a positive definite function. Differentiating $V(\mathbf{e}, \dot{\mathbf{e}})$ with respect to time yields

$$\dot{V}(\mathbf{e}, \dot{\mathbf{e}}) = \dot{\mathbf{e}}^T\ddot{\mathbf{e}} + \mathbf{e}^T\mathbf{K}_p(\mathbf{e})\dot{\mathbf{e}} \tag{39}$$

Combine the control law (32) and the dynamic model (10) and consider $\mathbf{S}^T\boldsymbol{\tau}_a = \boldsymbol{\tau}_N$, the closed-loop system equation can be written as

$$\mathbf{M}_e\left(\ddot{\mathbf{e}} + \mathbf{K}_p(\cdot)\mathbf{e} + \mathbf{K}_d(\cdot)\dot{\mathbf{e}}\right) = 0 \tag{40}$$

Since $\mathbf{M}_e$ is uniform positive definite, then $\mathbf{M}_e^{-1}$ exists and bounded, thus the closed-loop system equation (40) can be written as

$$\ddot{\mathbf{e}} + \mathbf{K}_p(\cdot)\mathbf{e} + \mathbf{K}_d(\cdot)\dot{\mathbf{e}} = 0 \qquad (41)$$

Multiplying both sides of the above equation by $\dot{\mathbf{e}}^T$, and then substituting the resulting equation into (39) yields

$$\dot{V} = -\dot{\mathbf{e}}^T \mathbf{K}_d(\cdot)\dot{\mathbf{e}} \qquad (42)$$

As $\mathbf{K}_d(\cdot)$ is a symmetric, positive definite matrix, then $\dot{V}$ is a semi-negative definite matrix, thus the closed-loop system is stable. Considering the closed-loop equation (41) is autonomous system, and defining the region Ω as

$$\Omega = \left\{ \begin{bmatrix} \mathbf{e} \\ \dot{\mathbf{e}} \end{bmatrix} : \dot{V}(\mathbf{e},\dot{\mathbf{e}}) = 0 \right\} = \left\{ \begin{bmatrix} \mathbf{e} \\ \dot{\mathbf{e}} \end{bmatrix} = \begin{bmatrix} \mathbf{e} \\ 0 \end{bmatrix} \in \Re^4 \right\} \qquad (43)$$

Thus $\begin{bmatrix} \mathbf{e} \\ \dot{\mathbf{e}} \end{bmatrix} = \begin{bmatrix} 0 \\ 0 \end{bmatrix}$ is the largest invariant set of $\Omega = \left\{ \begin{bmatrix} \mathbf{e} \\ \dot{\mathbf{e}} \end{bmatrix} : \dot{V}(\mathbf{e},\dot{\mathbf{e}}) = 0 \right\}$, and constitutes an asymptotically stable equilibrium point. By using the LaSalle's theorem, one can get that the closed-loop system of the parallel manipulator is asymptotically stable.

4. Nonlinear friction model and identification

In this section, the friction compensation method based on a nonlinear friction model is developed for the parallel manipulator. This nonlinear friction model enables reconstruction of viscous, Coulomb, and Stribeck friction effects of the parallel manipulator. Identification experiments are carried out, and parameters in the nonlinear friction model are estimated by nonlinear optimization.

4.1 Nonlinear friction modeling
In order to reconstruct the nonlinear friction effect, the nonlinear friction model can be formulated as (Hensen et.al., 2000; Kostic et.al., 2004)

$$f(\dot{\theta}) = B_v\dot{\theta} + \sum_{k=1}^{3} f_k(1 - \frac{2}{1+e^{2\omega_k\dot{\theta}}}), \quad k = 1,2,3 \qquad (44)$$

where the first term represents the viscous friction and B_v is the viscous friction coefficient. The other terms model the Coulomb and Stribeck friction effects. The parameters f_k represent the magnitude of the Coulomb friction and the Stribeck curve. The parameters ω_k determine the slope in the approximation of the sigmoid function in the Coulomb friction and the Stribeck curve.

Obviously, the nonlinear friction model is an odd continuous function. Since $f(\dot{\theta})$ is clearly zero at $\dot{\theta} = 0$, the model does not capture the static friction. The friction model does not describe stiction, because the system will always slide for an applied force unequal to zero. The stiction regime will be approximated, if the slope of the function near $\dot{\theta} = 0$ is very steep. Then the model can still give acceptable simulation results, i.e., angular displacement during stiction is neglectable. On the other hand, a continuous friction function will facilitate the numerical solution if such a model is used in parameter identification.

In (44), there are three sigmoid functions. If more sigmoid functions are selected, the estimation accuracy with this model will be better, but the friction model will have more parameters and it will be more complicated. So for this nonlinear friction model, a suitable number of the sigmoid function is important. One can also analyze this problem with the neural network. The nonlinear terms in (44) can be constructed with a two layers neural network, i.e., one hidden layer and one output layer. Defining the weight matrices for the first and second layer as $\mathbf{W}_1$ and $\mathbf{W}_2$, the neural network output can be written as (Hensen et.al., 2000)

$$f'(\dot{\theta}) = \mathbf{W}_2^T \sum (\mathbf{W}_1 \dot{\theta} + b_1) + b_2 \tag{45}$$

where b_i represents the bias value for the neurons in the i-th layer and $\sum(\cdot)$ is a nonlinear operator with $\sum(x) = [\sigma(x_1) \quad \sigma(x_2) \quad \sigma(x_3)]^T$, the activation function $\sigma(x) = 1 - \dfrac{2}{1 + e^{2x}}$. From (44), the parameter b1 and b2 are both zero. The weight matrix for the first layer and the second layer can be written as $\mathbf{W}_1 = [\omega_1 \quad \omega_2 \quad \omega_3]^T$ and $\mathbf{W}_2 = [f_1 \quad f_2 \quad f_3]^T$ respectively. As we known, increasing the number of the hidden neurons, the approximation performance with the network will be better. However, too many hidden neurons will make the network more complicated and the training time may be longer. In practice, suitable number of the hidden neurons should be selected. In order to model the nonlinear friction of the parallel manipulator, two hidden neurons are enough, that is to say two sigmoid functions will be selected.

If the friction of the passive joints is neglected, according to (44), one can define the nonlinear friction model for the 2-DOF planar parallel manipulator as follows

$$f_{ai} = B_{vi}\dot{q}_{ai} + f_{1i}(1 - \frac{2}{1 + e^{2\omega_{1i}\dot{q}_{ai}}}) + f_{2i}(1 - \frac{2}{1 + e^{2\omega_{2i}\dot{q}_{ai}}}) + d_i, \quad i = 1,2,3 \tag{46}$$

where the first term represents the viscous friction and B_{vi} is the viscous friction coefficient of the ith active joint; d_i represents the zero drift of the motion control board; the remaining terms model the Coulomb and Stribeck friction effects of the ith active joint. The parameter f_{1i} and f_{2i} represent the magnitude of the Coulomb friction and the Stribeck curve. The parameters ω_{1i} and ω_{2i} determine the slope in the approximation of the sigmoid function in the Coulomb friction and the Stribeck curve.

4.2 Nonlinear friction identification
In the dynamic model Eq. (10), the dynamic parameters can be calculated directly, and only the parameters in the nonlinear friction model Eq. (46) need to be identified. In Eq. (10), the mass, length and joint angles all united into the standard units. The corresponding torque has the unit N.m. Since the commanded torque for the motion control board of the parallel manipulator is digital value, the proportion should be obtained between the torque of the unit N.m and the commanded digital value of the torque (Shang et.al., 2008). Defining the dynamic torque $(\mathbf{S}^T)^+(\mathbf{M}_e\ddot{q}_e + \mathbf{C}_e\dot{q}_e) = \mathbf{D}$ and the proportion is k, the dynamic model Eq. (10) can be rewritten as

$$\mathbf{D} \cdot k + \mathbf{f}_a = \mathbf{\tau}_a \tag{47}$$

Substituting the nonlinear friction model (46) into (47), one can define the optimization function J as follows

$$J = \sum_{j=1}^{N} \sum_{i=1}^{3} \left(\tau_{ai}^j - k(D_i^j) - (B_{vi}\dot{q}_{ai}^j + f_{1i}(1 - \frac{2}{1 + e^{2\omega_{1i}\dot{q}_{ai}^j}}) + f_{2i}(1 - \frac{2}{1 + e^{2\omega_{2i}\dot{q}_{ai}^j}}) + d_i) \right)^2 \tag{48}$$

where τ_{ai}^j and D_i^j represent the actuator torque and the dynamic torque of the ith active joint in the jth configuration respectively. And $\dot{q}_{ai}^j$ represents the velocity of the ith active joint in the jth configuration.

The parameters B_{vi}, f_{1i}, f_{2i}, ω_{1i}, ω_{2i}, d_i, and the proportion k, a total of 19 parameters, are selected as the optimization variables. These parameters will be estimated by making the optimization function J minimum. Parameter optimization procedures are programmed with Matlab, and the nonlinear optimization function *fmincon* finding a constrained minimum of a function of several variables is called for in Matlab. In order to use the *fmincon*, the first step is set the initial value, the lower limit value and the upper limit value of the 19 optimization variables. Then (48) is defined as the optimized function of *fmincon*. The third step is getting the variables τ_{ai}^j, D_i^j, and $\dot{q}_{ai}^j$ in (48). Next we will give the procedures about getting the variables τ_{ai}^j, D_i^j, and $\dot{q}_{ai}^j$ in our actual identification experiment.

In actual identification experiment, the end-effector of the parallel manipulator is driven to track a circular trajectory. The center coordinates of the circle are (0.29 , 0.25) and radius is 0.07 , the unit is meter, this circle motion is repeated clockwise for 15 times. The parallel manipulator is controlled by the PD controller in the task space, the actuator torque is also the control input, thus τ_{ai}^j is a variable known. The control input is selected in the null-space of the matrix $\mathbf{S}^T$, thus the actuator torque τ_{ai}^j in (48) is also in the null-space of the matrix $\mathbf{S}^T$. And this selection will make the control input used in the identification experiment minimum. For the parallel manipulator, only angles of the active joints can be measured directly by the absolute optical-electrical encoders. The angular velocity of the active joints is obtained by numerical differentiation of the active joint angles, and a low-pass filter is adopted to filter the angular velocity signal, then we will get the variable $\dot{q}_{ai}^j$. The angular acceleration of the joints is obtained by numerical differentiation of the filtered angular velocity. With the velocity and the acceleration of the joints, and considering the kinematics of the parallel manipulator, the actual velocity and acceleration of the end-effector can be obtained. Thus D_i^j can be calculated with these variables. With the actual values of the variable τ_{ai}^j, D_i^j, and $\dot{q}_{ai}^j$, the unknown parameters of the parallel manipulator are identified and results are shown in Table 1.

4.3 Coulomb + viscous friction identification

In order to compare with the nonlinear friction model, a common *Coulomb + viscous* friction model containing the viscous friction and Coulomb friction effect is established for the parallel manipulator. The friction model can be written as

$$f_{ai} = sign(\dot{q}_{ai})f_{ci} + f_{vi}\dot{q}_{ai} + d_i , \; i = 1,2.3 \tag{49}$$

where f_{ci} represents the Coulomb friction; f_{vi} represents the coefficient of the viscous friction; d_i represents the zero drift of the motion control board.

parameters	values	parameters	values	parameters	values	parameters	values
k	508.7	ω_{21}	-16.4	ω_{12}	-18	f_{23}	500
B_{v1}	892.8	d_1	-2.6	ω_{22}	5.1	ω_{13}	-0.9
f_{11}	1040.1	B_{v2}	1396.4	d_2	-21.9	ω_{23}	10
f_{21}	-268.0	f_{12}	-309.0	B_{v3}	402.6	d_3	30
ω_{11}	0.7	f_{22}	-61.8	f_{13}	-867.7		

Table 1. Identification results of the nonlinear friction model.

With the analysis of the identification of the nonlinear friction model, the corresponding work of the *Coulomb + viscous* friction model is much simpler. Substituting the *Coulomb + viscous* friction model (49) into (47), one can get a linear equation about the identified parameters as follows

$$[\mathbf{D}\ \ \mathbf{K}]\begin{bmatrix} k & f_{v1} & f_{v2} & f_{v3} & d_1+f_{c1} & d_1-f_{c1} & d_2+f_{c2} & d_2-f_{c2} & d_3+f_{c3} & d_3-f_{c3} \end{bmatrix}^T = \tau_a \quad (50)$$

where

$$\mathbf{K} = \begin{bmatrix} \dot{q}_{a1} & 0 & 0 & u_1 & l_1 & 0 & 0 & 0 & 0 \\ 0 & \dot{q}_{a2} & 0 & 0 & 0 & u_2 & l_2 & 0 & 0 \\ 0 & 0 & \dot{q}_{a3} & 0 & 0 & 0 & 0 & u_3 & l_3 \end{bmatrix}$$

For simplicity, parameter combinations d_i+f_{ci} and d_i-f_{ci} are viewed as identified parameters, and the coefficients u_i and l_i of the parameters are determined by the following rules: $u_i = 1, l_i = 0$ when $\dot{q}_{ai} \geq 0$, and $u_i = 0, l_i = 1$ when $\dot{q}_{ai} < 0$.
There are 10 parameters to be identified in Eq. (50), but only three independent equations can be got for each sampling point. So a group of linear equations about the unknown parameters can be got with the sampling data of a continuous trajectory, then the Least Squares method is used to identify the unknown parameters.
The identification experiment designed for the *Coulomb + viscous* friction model is the same with the nonlinear friction model discussed in section 4.2. Identification results of the *Coulomb + viscous* friction model are shown in Table 2.

parameters	values	parameters	values
k	512.7	d_1-f_{c1}	-261.7
f_{v1}	1534.8	d_2+f_{c2}	212.6
f_{v2}	1415.9	d_2-f_{c2}	-256
f_{v3}	1475.1	d_3+f_{c3}	179.2
d_1+f_{c1}	248.5	d_3-f_{c3}	-129

Table 2. Identification results of the *Coulomb + viscous* friction model

5. Experiments

As shown in Fig. 2, the actual experiment platform is a 2-DOF parallel manipulator with redundant actuation designed by Googol Tech. Ltd. in Shenzhen, China. It is equipped with three permanent magnet synchronous servo motors with harmonic gear drives. The active joint angles are measured with absolute optical-electrical encoders. The nonlinear dynamic controllers and the friction compensation method are programmed with the Visual C++, and the algorithms run on a Pentium III CPU at 733MHz. with the sampling period 2ms.

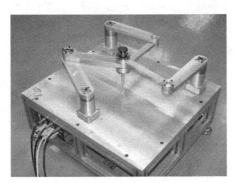

Fig. 2. The prototype of the 2-DOF parallel manipulator with redundant actuation

5.1 Experiments of the ANPD controller

The trajectory tracking control experiment is designed for the parallel manipulator to validate the ANPD controller. The desired trajectory of the end-effector is a straight line, the starting point is (0.22, 0.29) and the ending point is (0.37, 0.21), thus the motion distance is 0.17m. The profile of the desired velocity is an S-type curve (Cheng et.al., 2003). In the experiment, the low-speed and high-speed motions are both tested. For the low-speed motion, the max velocity is 0.2m/s, the max acceleration is 5m/s², and the jerk is 200m/s³. For the high-speed motion, the max velocity is 0.5m/s, the max acceleration is 10m/s², and the jerk is 400m/s³.

In order to implement the ANPD controller (17), the dynamic parameters in (18.a) and the friction parameters in (18.b) must be known. In the experiment, the nominal values of the dynamic parameters are used (Shang et.al., 2008). Then, with the known dynamic parameters, the friction parameters in the *Coulomb + viscous* friction model can be identified by the Least Squares method, as shown in Table 2. In fact, the control parameters in (18.c) are tuned and determined by the actual experiments. The procedures to tune the control parameters in (18.c) can be summarized as follows:

1. Assume $k_{p1} = k_{p2} = k_p$, $k_{d1} = k_{d2} = k_d$. Let $k_d = 0$, $\alpha_1 = 1$, $\alpha_2 = 1$, and increase the value of k_p from zero until the system show a little oscillation to some extent.
2. Keep the value of k_p tuned well in the first stage, and increase the value of k_d to improve the dynamic performance further.
3. Regulate finely the above two values and make tradeoffs between k_p and k_d.
4. Find the maximum error and error rate of the end-effector under the tuned value of k_p and k_d.
5. In the ANPD controller, δ_1 and δ_2 are the threshold of the error and the error rate. If δ_1 is tuned bigger than the maximum error, then the proportional gain $k_p(e_i)$ will always equals to $k_p \delta_1^{\alpha_1 - 1}$; and δ_1 is tuned close to 0, then $k_p(e_i)$ will always equal to $k_p |e_i|^{\alpha_1 - 1}$. So, δ_1 should be made a tradeoff between the maximum error and 0 error. Similar method can be used to tune parameter δ_2. From our actual experiences, the

value of δ_1 is tuned to the half value of the maximum error, and the value of δ_2 is tuned to the half of the maximum error rate. This choice has good control performance and it's easy to implement.

6. For the parameters $\alpha_1 = 1$ and $\alpha_2 = 1$, the proportional gain $k_p(e_i)$ is a constant of k_p, and the derivative gains $k_d(\dot{e}_i)$ is a constant of k_d. Thus the NPD algorithm can be considered as the linear PD algorithm. So decrease the value of α_1 ($0.5 \leq \alpha_1 \leq 1$), and decrease the value of k_p at the same time to improve the error curve further, and make tradeoffs between the two values. Using this step, one can get the nonlinear proportional gain of the ANPD controller.

7. Increase the value of α_2 ($1 \leq \alpha_2 \leq 1.5$), and decrease the value of k_d at the same time to improve the error rate curve further, then make tradeoffs between the two values.

Using the above procedures, the ANPD controller parameters are tuned as follow:

$$k_p = 4500 \,, k_d = 470 \,, \delta_1 = 3 \times 10^{-4} \,, \delta_2 = 3 \times 10^{-3} \,, \alpha_1 = 0.7 \,, \alpha_2 = 1.1$$

In order to make a comparison between the ANPD controller and the APD controller, the same tracking experiments are implemented on the parallel manipulator. We choose the APD controller is because it has nonlinear dynamics compensation and friction compensation. In the APD controller, the control input vector of the three actuated joints can be calculated as (Shang et.al., 2009)

$$\tau_a = (S^T)^+ \left(M_e \ddot{q}_e^d + C_e \dot{q}_e^d + K_{lp} e + K_{ld} \dot{e} \right) + f_a \tag{51}$$

where K_{lp} and K_{ld} are both symmetric, positive definite matrices of constant gains. In the APD controller (51), M_e and C_e can be calculated with the nominal dynamic parameters, and f_a can be calculated with the values of the friction parameters shown in Table 2. The procedures of tuning parameters K_{lp} and K_{ld} in APD controller are similar to the procedures of tuning parameters k_p and k_d in ANPD controller. Thus, the tuning procedures (1) to (3) can be used to tune the parameters K_{lp} and K_{ld}.

The experiment results of the APD and ANPD controller are shown in Fig. 3-4. Fig. 3a and Fig. 3b are the tracking errors of the end-effector at the low-speed on the X-direction and Y-direction respectively. From the experimental curves, one can see that the ANPD controller can decrease the tracking errors during the whole motion process obviously, and the maximum error in the motion is smaller. Fig. 4a and Fig. 4b are the tracking errors of the end-effector at the high-speed on the X-direction and Y-direction respectively. One can find that the tracking errors are much smaller with the ANPD controller than with the APD controller, especially at the acceleration process. And one can conclude that, by the ANPD controller, the performance improvement of trajectory tracking accuracy at the high-speed is more obvious than at the low-speed.

Furthermore, to evaluate the performances of the two controllers, the root-square mean error (RSME) of the end-effector position is selected as the performance index

$$RSME = \sqrt{\frac{1}{N} \sum_{j=1}^{N} \left(e_x^2(j) + e_y^2(j) \right)}$$

$$= \sqrt{\frac{1}{N} \sum_{j=1}^{N} \left(\left(x^d(j) - x(j) \right)^2 + \left(y^d(j) - y(j) \right)^2 \right)} \tag{52}$$

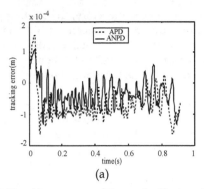 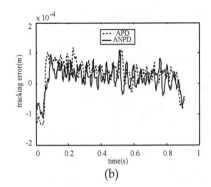

Fig. 3. Tracking errors of the end-effector at the low-speed: (a) X-direction; (b) Y-direction

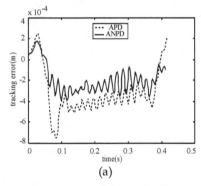 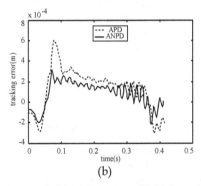

Fig. 4. Tracking errors of the end-effector at the high-speed: (a) X-direction; (b) Y-direction.

where $x^d(j)$ and $y^d(j)$ represent the X-direction and Y-direction position coordinates at the jth sampling point of the desired trajectory respectively; $x(j)$ and $y(j)$ represent the X-direction and Y-direction position coordinates of the jth sampling point of the actual trajectory respectively.

The RSME results of the trajectory tracking experiment of the ANPD and APD controller are shown in Table 3. From the data of the RPE (reduced percentage of error) in Table 3, the ANPD controller can increase the position accuracy of the end-effector above 30%, compared with the conventional APD controller.

	at slow-speed(m)	at high-speed(m)
APD	1.04×10^{-4}	4.55×10^{-4}
ANPD	7.22×10^{-5}	2.78×10^{-4}
RPE	30.6%	38.9%

Table 3. RSME of the APD and ANPD controller

5.2 Experiments of the NCT controller

In order to validate the NCT controller further, the trajectory tracking control experiment is designed for the parallel manipulator. Both the linear and circular trajectories in the workspace are selected as the desired trajectory. For the linear trajectory, the starting point is (0.22, 0.19) and the ending point is (0.35, 0.29), thus the motion distance is 0.164m. The velocity profile of the linear trajectory is an S-type curve (Cheng et.al., 2003), the max velocity is 0.5m/s, the max acceleration is 10m/s², and the jerk is 400m/s³. For the circular trajectory with the constant speed of 0.5m/s, the center is (0.29, 0.25) and the starting point is (0.29, 0.31), thus the radius is 0.06m.

The actual implement of the NCT controller is similar to the ANPD controller, and the dynamic parameters in (33.a) and the friction parameters in (33.b) must be known. In the experiment, the nominal values are selected as the values of the actual dynamic parameters (Shang et.al., 2008). Then, with the known dynamic parameters, the friction parameters can be identified by the Least Squares method (Shang et.al., 2008). And the values of the control parameters in (33.c) are tuned and determined by the actual experiments. The tuning procedures for the ANPD controller can be used to tune the NCT controller. Using those procedures, the NCT controller parameters are tuned as follows: $k_p = 2400$, $k_d = 240$, $\delta_1 = 3 \times 10^{-4}$, $\delta_2 = 3 \times 10^{-3}$, $\alpha_1 = 0.7$, $\alpha_2 = 1.1$. Moreover, to demonstrate that the NCT controller can improve the tracking accuracy of the end-effector, experiments using the CT controller are carried out as comparison (Shang & Cong, 2009). The CT controller is chosen because it has friction compensation and feedback dynamics compensation. In the CT controller, the control input vector of the three active joints can be calculated as

$$\tau_a = \left(\mathbf{S}^T\right)^+ \left(\mathbf{M}_e \ddot{\mathbf{q}}_e^d + \mathbf{C}_e \dot{\mathbf{q}}_e + \mathbf{M}_e \left(\mathbf{K}_{lp} e + \mathbf{K}_{ld} \dot{e}\right)\right) + \mathbf{f}_a \qquad (53)$$

where $\mathbf{K}_{lp}$ and $\mathbf{K}_{ld}$ are both symmetric, positive definite matrices of constant gains.

In the CT controller (53), the dynamic parameters in $\mathbf{M}_e$ and $\mathbf{C}_e$, and the friction parameters in $\mathbf{f}_a$ are the same with these of the NCT controller. The procedures of tuning parameters of $\mathbf{K}_{lp}$ and $\mathbf{K}_{ld}$ in the CT controller are similar to the procedures of tuning parameters of k_p and k_d in the NCT controller. Thus, the tuning procedures (1), (2), and (3) can be used to tune the parameters of $\mathbf{K}_{lp}$ and $\mathbf{K}_{ld}$. Using the above methods, the CT controller parameters are tuned as follows: $\mathbf{K}_{lp} = diag(20000, \ 20000)$, $\mathbf{K}_{ld} = diag(150, \ 150)$.

The tracking error curves of the end-effector controlled by the CT and NCT controller are shown in Fig. 5-6. Fig. 5 is the linear trajectory tracking errors of the end-effector on the X-direction and Y-direction. From the experiment curves, one can see that the NCT controller can decrease the tracking errors during the whole motion process obviously, and the maximum error in the motion is smaller. Fig. 6 is the circular trajectory tracking errors of the end-effector on the X-direction and Y-direction. From the curves one can see, the tracking accuracy is improved obviously using the NCT controller, compared with the CT controller. The RSME results of the trajectory tracking experiment of the NCT and CT controller are shown in Table 4. From the data of the RPE in Table 4, the NCT controller can increase the position accuracy of the end-effector above 35%, compared with the conventional CT controller.

	Line (m)	Circle (m)
CT	4.77×10⁻⁴	4.41×10⁻⁴
NCT	3.08×10⁻⁴	2.59×10⁻⁴
RPE	35.4%	41.3%

Table 4. RSME of the CT and NCT controller

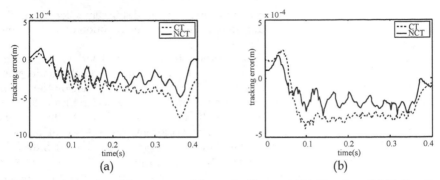

Fig. 5. Linear trajectory tracking errors of the end-effector: (a) X-direction; (b) Y-direction

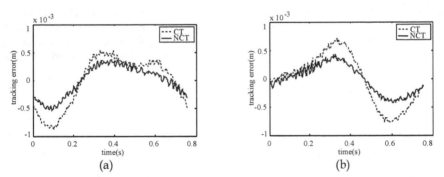

Fig. 6. Circular trajectory tracking errors of the end-effector: (a) X-direction; (b) Y-direction.

5.3 Experiments of nonlinear friction compensation

In order to compare with the compensation performances of the nonlinear friction model and the Coulomb + viscous friction model, the trajectory tracking experiments are implemented on the parallel manipulator. In the actual experiment, the augmented PD (APD) controller is designed in the task space for the parallel manipulator (Shang et.al., 2009). In the APD controller, the control input vector of the three active joints can be calculated as

$$\boldsymbol{\tau}_a = (\mathbf{S}^T)^+ \left(\mathbf{M}_e \ddot{\mathbf{q}}_e^d + \mathbf{C}_e \dot{\mathbf{q}}_e^d + \mathbf{K}_{lp}\mathbf{e} + \mathbf{K}_{ld}\dot{\mathbf{e}} \right) + \mathbf{f}_a \tag{54}$$

where the term $\mathbf{f}_a$ is the friction compensation calculated by the nonlinear friction model (46) with the parameter values in Table 1. Moreover, $\mathbf{f}_a$ can be calculated by the Coulomb + viscous friction model (49) with the parameter values in Table 2. If the term $\mathbf{f}_a$ is neglected in the APD controller (54), it means the friction compensation is not considered in the controller and the friction is ignored in the parallel manipulator.

Both the straight line and the circle in the task space are selected as the desired trajectory to study the friction compensation. For the straight line, the starting point is (0.22, 0.29) and the ending point is (0.37, 0.21), thus the motion distance is 0.17m. The profile of the desired velocity is trapezoidal curve. In the experiment, both the low-speed and high-speed motions are implemented. For the low-speed motion, the maximum velocity is 0.2m/s and the acceleration is 5m/s². For the high-speed motion, the maximum velocity is 0.5m/s and the acceleration is 10m/s². In the circle motion with constant speed, the center coordinates of the circle are (0.29, 0.25) and radius is 0.04, also both the low-speed motion of 0.2m/s, and the high-speed motion of 0.5m/s are implemented.

Linear trajectory tracking errors of the end-effector at the slow-speed and the high-speed are shown in Fig.7. From the curves one can see, the tracking errors are much smaller with the friction compensation methods based on the *Coulomb + viscous* model or the nonlinear model, compared with the without friction compensation method which means the term f_a is neglected in the APD controller (54). Especially, the maximum error at the acceleration process is decreased greatly with the friction compensation methods. Also one can see that, compared with the friction compensation based on the *Coulomb + viscous* model, the tracking accuracy has improved further at the low-speed, and the improvement at the high-speed is even more apparent by using the friction compensation based on the nonlinear model.

Circular trajectory tracking errors of the end-effector at the slow-speed and the high-speed are shown in Fig.8. From the curves one can see, the tracking accuracy is improved obviously using the two friction compensation methods, compared with the without friction compensation method. Also one can see that, with the friction compensation based on the nonlinear model, the tracking error is decreased at the low-speed, and the improvement at the high-speed is more obvious than the friction compensation based on the *Coulomb + viscous*.

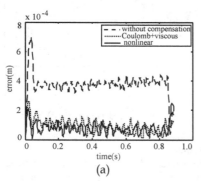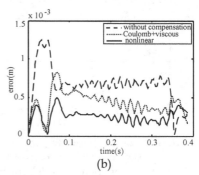

Fig. 7. Linear trajectory tracking error of the end-effector: (a) at the low-speed; (b) at the high-speed.

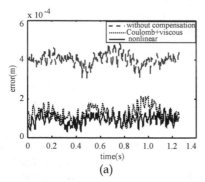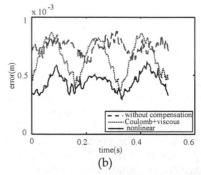

Fig. 8. Circular trajectory tracking error of the end-effector: (a) at the low-speed; (b) at the high-speed

Furthermore, the RSMEs of the trajectory tracking experiment are shown in Table 5. From the data in the table one can see, by using the two friction compensation methods, the RSMEs are much smaller than the method ignoring friction compensation. And the RSMEs of the friction compensation based on the nonlinear model are smaller than the friction compensation with the *Coulomb + viscous* model, especially when the speed is higher.

Friction compensation method	Straight line motion (m)		Circle motion (m)	
	0.2m/s	0.5m/s	0.2m/s	0.5m/s
Without compensation	3.83×10^{-4}	7.07×10^{-4}	4.00×10^{-4}	7.45×10^{-4}
Coulomb + viscous model	1.03×10^{-4}	4.56×10^{-4}	1.29×10^{-4}	6.68×10^{-4}
Nonlinear model	8.88×10^{-5}	2.71×10^{-4}	9.53×10^{-5}	4.49×10^{-4}

Table 5. RSME of the trajectory tracking experiments

6. Conclusions

In order to realize the high-speed and high-accuracy motion control of parallel manipulator, nonlinear control method is used to improve the traditional dynamic controllers such as the APD controller and the CT controller. The common feature of the two controllers is eliminating tracking error by linear PD control, and the friction compensation is realized by using the Coulomb + viscous friction model. However, the linear PD control is not robust against the uncertain factors such as modeling error and external disturbance. To overcome this problem, the NPD control is combined with the conventional control strategies and two nonlinear dynamic controllers are developed. Moreover, a nonlinear model is used to construct the friction of the parallel manipulator, and the nonlinear friction can be compensated effectively. Our theory analysis implies that, the proposed controllers can guarantee asymptotic convergence to zero of both tracking error and error rate. And for its simple structures and design, the proposed controllers are easy to be realized for the industry applications of parallel manipulators. Our experiment results show that, the position error of the end-effector decrease obviously with the proposed controllers and the nonlinear friction compensation method, especially at the high-speed. So the nonlinear dynamic controller and nonlinear friction compensation can realize high-speed and high-accuracy trajectory tracking of the parallel manipulator in practice. Also these new methods can be used to other manipulators, such as serial ones, or parallel manipulator without redundant actuation to realize high-speed and high accuracy motion.

7. Acknowledgments

This work was supported by the National Natural Science Foundation of China with Grant No. 50905172, the Anhui Provincial Natural Science Foundation with Grant No.090412040, and the Fundamental Research Funds for the Central Universities.

8. References

Cheng H., Yiu Y.K., Li Z.X. (2003) Dynamics and control of redundantly actuated parallel manipulators. *IEEE Trans. Mechatronics*, 8(4): 483-491

Ghorbel F.H., Chetelat O., Gunawardana R., Longchamp R. (2000) Modeling and set point control of closed-chain mechanisms: theory and experiment. *IEEE Trans. Control Syst. Tech.*, 8(5):801-815

Hensen R.H.A., Angelis G.Z., Molengraft M.J.G., Jager A.G., Kok J.J. (2000) Grey-box modeling of friction: An experimental case-study. *European Journal of Control* 6(3):258-267

Han J.Q. (1994) Nonlinear PID controller. *Acta Automatica Sinica*, 20(4): 487-490.

Kelly R., Ricardo C. (1996) A class of nonlinear PD-type controller for robot manipulator. *Journal of Robotic Systems*, 13: 793-802

Kostic D., Jager B., Steinbuch M., Hensen R. (2004) Modeling and identification for high-performance robot control: an RRR-robotic arm case study. *IEEE Trans. Contr. Syst. Techn.* 12(6): 904-919

Li Q., Wu F.X. (2004) Control performance improvement of a parallel robot via the design for control approach. *Mechatronics*, 14(8): 947-964

Merlet J.P. (2000) *Parallel robots*. Norwell, MA: Kluwer

Muller A. (2005) Internal preload control of redundantly actuated parallel manipulators - Its application to backlash avoiding control, *IEEE Trans. Robotics*, 21(4), 668 - 677

Murray R., Li Z. X., Sastry S. (1994) A Mathematical Introduction to Robotic Manipulation. CRC Press

Ouyang P.R., Zhang W.J., Wu F.X. (2002) Nonlinear PD control for trajectory tracking with consideration of the design for control methodology. In: *Proc. of the IEEE Int. Conf. Robot. Autom.*, Washington; May, 2002, pp. 4126-4131

Paccot F., Andreff N., Martinet P. (2009) A review on the dynamic control of parallel kinematic machines: theory and experiments. *International Journal of Robotics Research*, 28(3): 395-416

Seraji H. (1998) A new class of nonlinear PID controllers with robotic applications. *Journal of Robotic Systems*, 15(3): 161-181

Shang W.W., Cong S., Zhang Y.X. (2008) Nonlinear friction compensation of a 2-DOF planar parallel manipulator. *Mechatronics*, 18(7): 340-346

Shang W. W., Cong S. (2009) Nonlinear computed torque control for a high speed planar parallel manipulator, *Mechatronics*, 19(6): 987-992

Shang W. W., Cong S., Li Z. X., Jiang S. L. (2009) Augmented nonlinear PD controller for a redundantly actuated parallel manipulator. *Advanced Robotics*, 23: (12-13), 1725-1742

Shang W. W., Cong S., Kong F. R. (2010) Identification of dynamic and friction parameters of a parallel manipulator with actuation redundancy, *Mechatronics*, 20(2): 192-200.

Shang W. W., Cong S., Jiang S. L. (2010) Dynamic model based nonlinear tracking control of a planar parallel manipulator, *Nonlinear Dynamics*, 60(4): 597-606

Slotine J. -J. E., Li W. P. (1991) *Applied nonlinear control*, Englewood Cliffs, N. J.: Prentice Hall.

Su Y.X., Duan B.Y., Zheng C.H., Zhang Y.F. Chen G.D., Mi J.W. (2004) Disturbance-rejection high-precision motion control of a Stewart platform. *IEEE Trans. Control Syst. Tech.*, 12(3): 364-374

Su Y.X., Zheng C.H., Duan B.Y. (2005) Fuzzy learning tracking of a parallel cable manipulator for the square kilometer array. *Mechatronics*, 15(6):731-746

Wu F.X., Zhang W.J., Li Q., Ouyang P.R. (2002) Integrated design and PD control of high-speed closed-loop mechanisms. *J. Dyn. Syst., Meas., Control*, 124(4): 522-528

Xu Y., Hollerbach J.M., Ma D. (1995) A nonlinear PD controller for force and contact transient control. *IEEE Control Systems Magazine*, 15: 15-21.

Zhang Y.X., Cong S., Shang W.W., Li Z.X., and Jiang S.L. (2007) Modeling, identification and control of a redundant planar 2-Dof parallel manipulator. *Int. J. Control, Autom. Syst.*, 5(5):559-569

2

Brushless Permanent Magnet Servomotors

Metin Aydin

Kocaeli University, Department of Mechatronics Engineering, Kocaeli
Turkey

1. Introduction

Electrical motors drive a variety of loads in today's world. Almost every industrial process relies on some kind of electrical motors and generators. There exist billions electric motors used in different applications all over the world. Majority of them are small fractional HP motors use in household appliances. However, they used about 5% of the electricity used by the motors. Three phase motors are used in heavier applications and consume substantial amount of electricity. These electric motors operate long hours and consume more than half of the electricity used by motors.

The oldest type of electric motor, wound field DC motor, was the most popular motor for years and easiest for speed control. Although they are replaced by adjustable AC drives in many applications, they are still used in some low power and cost effective applications. The main reason why DC drives faded away over the last decade is that they require converters and maintenance, not to mention their lower torque densities compared to AC motors. Induction motors are also one of the most widely used motors in AC drive applications. They are reliable and don't require maintenance due to the absence of brushes and slip rings. The availability of single phase power is another big plus for these motors. The fact that the rotor windings are present makes the induction motors less efficient and creates cooling problems of the rotor. One crucial drawback of the induction motors is the parameter variation due to the heat caused by the rotor winding.

Variable reluctance motors are also frequently used in the industry and robotics. It's simple and robust stator and rotor structures reduce the cost dramatically compared to other types of motors. The converter requirement is also not very severe. A simple half bridge converter can easily be used to drive the motor. On the other hand, variation of reluctance does also create significant cogging, vibration and audible noise.

As for the synchronous motors, they have benefits and drawbacks of both DC and induction motors. The synchronous motors with field winding can be more efficient than a DC or induction motors and are used in relatively large loads such as generating electricity in power plants. If the rotor winding in synchronous motors is replaced by permanent magnets, another variation of synchronous motors is obtained. These motors are called permanent magnet motors which can be supplied by sinusoidal or trapezoidal currents. These motors have three major types based on their magnet structures as displayed in Fig. 1.

The lack of slip rings and rotor windings as well as high power density, high efficiency and small size make these motors very attractive in the industrial and servo applications. In

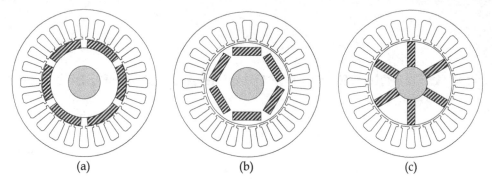

(a)   (b)   (c)

Fig. 1. Surface mounted PM (a), buried PM (b) and spoke type PM (c) motor types

addition, PM servomotors have better torque-speed characteristics and high dynamic response than other motors. Their long operating lives, noise-free operations and high speed ranges are some of the advantages of brushless servomotors.

2. Classification of electric motors

2.1 General motor classification

There are several ways of classifying electric motors by their electrical supply, by rotor structures and stator types. One of the common ways is to categorize them as AC and DC motors as shown in Fig. 2. AC motors use alternating current or voltage as source while DC motors use DC voltage source to supply the windings. DC motors are classified by their field connections such as series, parallel or compound field excitation. AC motors, on the other hand, has two major types: One type is induction motors where rotor magnetic field is generated by electromagnetic induction principles and the other is synchronous motors where the magnetic field is generated by either field winding excitation or permanent magnets. Induction motors could be single or poly-phase and have squirrel-cage or wound rotor. Synchronous motors could have numerous options depending on the rotor type and excitation (Hendershot & Miller, 1995).

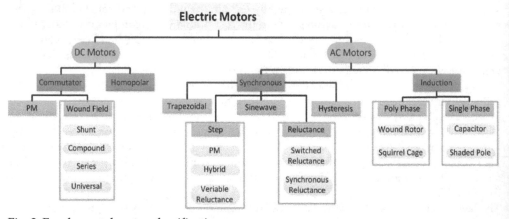

Fig. 2. Fundamental motor classification

Electric motors are also classified by their slots. They are called slotted motors if they do have slots and called non-slotted or slotless motors if they do not have any slot structures. Furthermore, one major classification method is identified by the main flux direction. If the motor has a main flux component which is radial to the shaft, they are called radial flux motors and if the flux component is axial to the motor shaft, then the motors are called axial flux motors where they find various applications because of their structural flexibility.

2.2 Permanent magnet servomotors

There exist various permanent magnet (PM) servomotors in the literature. They can be classified into two main categories, which are surface mounted PM motors where magnets are glued on the rotor surface and buried PM motors where magnets are buried into the rotor.

The use of surface mounted PM motors increases the amount of PM material per pole used in the motor. Using more magnet material usually increases the torque production of the motor while it also increases the motor volume and thus the cost. Buried PM motor and interior PM motor use the flux concentration principles where the magnet flux is concentrated in the rotor core before it gets into the airgap. These motors usually have considerable reluctance torque which arises from the fact that the use of flux concentration in the iron core introduces a position dependent inductance and hence reluctance torque that can be beneficial in certain cases.

PM motors are also classified based on the flux density distribution and the shape of the current excitation. They are listed into two categories, one of which is PM synchronous motors (PMSM) and the other is PM brushless motors (BLDC). PMSM, also called permanent magnet AC (PMAC) motors, has sinusoidal flux density, current and back EMF variation while the BLDC has rectangular shaped flux density, current variation and back EMF. Classification of these two motor types is explained in Table 1.

	PMSM	BLDC
Phase current excitation	Sinusoidal	Trapezoidal
Flux density	Sinusoidal	Square
Phase back EMF	Sinusoidal	Trapezoidal
Power and Torque	Constant	Constant

Table 1. Classification of permanent magnet motors based on their excitation and back EMF waveforms

	Surface PM motor	Buried/Interior PM motor
Convenience	BLDC	PMSM
Flux distribution	Square or Sinusodial	Usually Sinusoidal
Complexity of rotor	Simple	Complex
Speed limit	~1.2 x ω_R	~3 x ω_R or higher
High speed capability	Difficult	Possible
Control	Relatively easy	More complex

Table 2. Basic comparison of surface magnet and buried magnet motors

Fig. 3. Typical servomotor (Courtesy of FEMSAN Motor Co.)

Each PM motor type explained has some advantages over another. For instance, surface magnet motor has very simple rotor structure with fairly small speed limits. Buried or interior PM motors have wide speed ranges but their rotor is more complex than both surface magnet and inset PM rotors. In addition, buried or interior PM motors can go up to very high speeds unlike surface magnet motors although their control is more complex than surface magnet type motors. This comparison is also tabulated in Table 2.

2.3 Permanent magnet servomotor structure

A conventional surface mounted PM servomotor structure is illustrated in Fig. 1 (a). The motor has a stator and a PM rotor. The stator structure is slotted and formed by the laminated magnetic steel. A close picture of a laminated stator is shown in Fig. 4. Polyphase windings are placed into the stator slots although a slotless versions of servomotors are also available. The rotor structure is formed by the permanent magnets mounted on the rotor surface, rotor core and shaft. The rotor core is usually laminated. Fig. 5 shows both the stator and the rotor of a typical permanent magnet servomotor with high energy NdFeB magnets.

Fig. 4. Stator stack showing the servomotor laminations

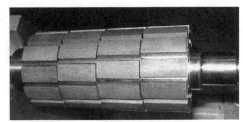

Fig. 5. Stator stack showing the motor laminations

2.4 PM servomotor torque-speed and back-EMF characteristics

Fig. 6 shows typical torque-speed characteristics of a brushless PM servomotor. There are two main torque parameters to describe a PM servomotor: Rated torque (T_R) and maximum torque (T_{max}). In addition, there are two major speed points: Rated speed and maximum speed. The region up to rated speed is called constant torque region and the region between the max speed (ω_{max}) and rated speed (ω_R) is called constant power region. During constant torque region, the motor can be loaded up to rated torque usually without any thermal problem. On the other hand, during constant power region, the motor torque starts to drop but the power stays almost constant. Another important characteristic of a PM motor is maximum load point which shows the overload capability of the motor. During this period, the motor can deliver higher torque for a short time to handle cases such as motor overload, start-up etc.

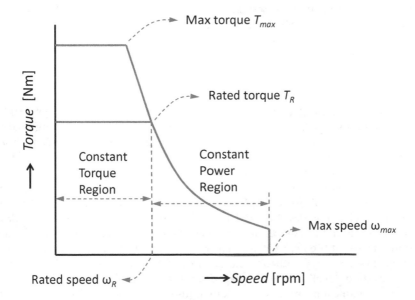

Fig. 6. Torque-speed characteristics of a PM servomotor

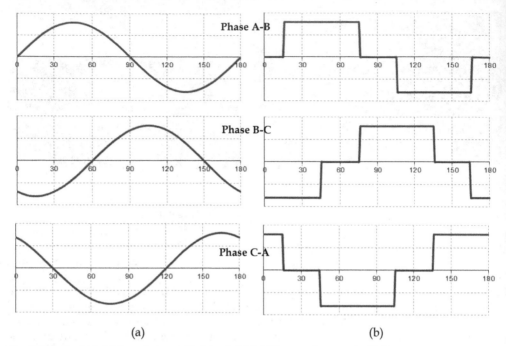

(a)   (b)

Fig. 7. Trapeziodal (a) and sinusoidal back-EMF (b) waveforms of a PM servomotor

There are two types of PM servomotor alternatives: Sinusoidal and trapezoidal motors. This is made on the basis of back-EMF waveforms. Trapezoidal servomotors have a back-EMF in trapezoidal manner and sinusoidal servomotors have a sinusoidal back-EMF as illustrated in Fig. 7. In addition to back-EMF, the supply current is trapezoidal and sinusoidal in each individual type of motors.

3. Magnetic materials

3.1 Magnetic steel
There exist various electric steel materials used in servomotors. Material type and grade depends mainly on the application and cost. High quality materials with high saturation and low loss levels are used in high performance and high speed applications while thick and high loss materials are used in low speed and cost effective applications. Non-oriented electrical steels are usually used in electric motor applications. Low magnetic loss and high permeability characteristics are valuable for applications where energy efficient, low loss, low noise and small size are important. One of the most frequently used magnetic steel lamination material is M270-35A (similar to M19 in the US). This material or similar grade is used in most PM servomotor applications. If high saturation levels and low losses at high speeds are required, materials such as Vacoflux50 would be a good option. The BH curve of these materials in addition to materials with high loss and thin high saturation level are all displayed in Fig. 8. Moreover, Table 3 shows the electrical and mechanical properties of various non-oriented electrical steel materials used in different motor applications.

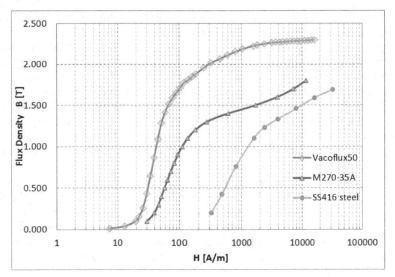

Fig. 8. Examples of steel materials with magnetic and structural properties

Grade EN 10106	Thickness mm	Maximum specific total loss at 50 Hz J=1.5 T W/kg	1.0 T W/kg	Minimum magnetic polarization at 50 Hz H=2500 T	5000 T	10000 A/m T
M235-35A	0.35	2.35	0.95	1.49	1.60	1.70
M250-35A	0.35	2.50	1.00	1.49	1.60	1.70
M270-35A	0.35	2.70	1.10	1.49	1.60	1.70
M300-35A	0.35	3.00	1.20	1.49	1.60	1.70
M330-35A	0.35	3.30	1.30	1.49	1.60	1.70
M700-35A*	0.35	7.00	3.00	1.60	1.69	1.77
M250-50A	0.50	2.50	1.05	1.49	1.60	1.70
M270-50A	0.50	2.70	1.10	1.49	1.60	1.70
M290-50A	0.50	2.90	1.15	1.49	1.60	1.70
M310-50A	0.50	3.10	1.25	1.49	1.60	1.70
M330-50A	0.50	3.30	1.35	1.49	1.60	1.70
M350-50A	0.50	3.50	1.50	1.50	1.60	1.70
M400-50A	0.50	4.00	1.70	1.53	1.63	1.73
M470-50A	0.50	4.70	2.00	1.54	1.64	1.74
M630-50A	0.50	5.30	2.30	1.56	1.65	1.75
M600-50A	0.50	6.00	2.60	1.57	1.66	1.76
M700-50A	0.50	7.00	3.00	1.60	1.69	1.77
M800-50A	0.50	8.00	3.60	1.60	1.70	1.78
M940-50A	0.50	9.40	4.20	1.62	1.72	1.81
M310-65A	0.65	3.10	1.25	1.49	1.60	1.70
M330-65A	0.65	3.30	1.35	1.49	1.60	1.70
M350-65A	0.65	3.50	1.50	1.49	1.60	1.70
M400-65A	0.65	4.00	1.70	1.52	1.62	1.72
M470-65A	0.65	4.70	2.00	1.53	1.63	1.73
M530-65A	0.65	5.30	2.30	1.54	1.64	1.74
M600-65A	0.65	6.00	2.60	1.56	1.66	1.76
M700-65A	0.65	7.00	3.00	1.57	1.67	1.76
M800-65A	0.65	8.00	3.60	1.60	1.70	1.78
M1000-65A	0.65	10.00	4.40	1.61	1.71	1.80

Grade EN 10106	Conventional density kg/dm³	Resistivity μΩcm	Yield strength N/mm²	Tensile strength N/mm²	Young's Modulus (E) RD N/mm²	TD N/mm²	Hardness HV5 (VPN)
M235-35A	7.60	59	460	580	185 000	200 000	220
M250-35A	7.60	55	455	575	185 000	200 000	215
M270-35A	7.65	52	450	565	185 000	200 000	215
M300-35A	7.65	50	376	490	185 000	200 000	185
M330-35A	7.65	44	300	430	200 000	220 000	150
M700-35A*	7.80	30	290	405	210 000	220 000	125
M250-50A	7.60	59	475	590	175 000	190 000	220
M270-50A	7.60	55	470	585	175 000	190 000	220
M290-50A	7.60	55	465	580	185 000	200 000	220
M310-50A	7.65	52	385	500	185 000	200 000	190
M330-50A	7.65	50	375	495	185 000	200 000	185
M350-50A	7.65	44	305	450	200 000	210 000	165
M400-50A	7.70	42	305	445	200 000	210 000	160
M470-50A	7.70	39	300	435	200 000	210 000	155
M530-50A	7.70	36	295	430	200 000	210 000	150
M600-50A	7.75	30	285	405	210 000	220 000	125
M700-50A	7.80	25	285	405	210 000	220 000	125
M800-50A	7.80	23	300	415	210 000	220 000	130
M940-50A	7.85	18	300	415	210 000	220 000	130
M310-65A	7.60	59	465	590	175 000	190 000	220
M330-65A	7.60	55	460	585	185 000	205 000	220
M350-65A	7.60	52	375	490	185 000	205 000	185
M400-65A	7.65	44	310	450	185 000	205 000	165
M470-65A	7.65	42	305	445	185 000	205 000	160
M530-65A	7.70	39	300	425	190 000	210 000	145
M600-65A	7.75	36	300	420	190 000	210 000	140
M700-65A	7.75	30	290	395	210 000	220 000	125
M800-65A	7.80	25	300	405	210 000	220 000	130
M1000-65A	7.80	18	295	400	210 000	220 000	125

Table 3. Non-oriented electric steel material properties (Source: Cogent)

3.2 Permanent magnets

Permanent magnet materials have been used in electric motors for decades. One important property of permanent magnets is the maximum energy product (MEP) which is the multiplication of residual flux density (B_r) and coercive force (H_r). In other words, MEP represents the maximum energy available per unit volume (kJ/m³). MEP is also an indication of magnet force. Furthermore, the larger the MEP, the smaller the magnet material needed for the same force. Permeability is another important property of the magnets. It is the slope of the demagnetization curve in the linear region. Small permeability means high flux levels before the magnet is irreversibly demagnetized.

Alnico magnets which are Aluminum, nickel, iron and later addition of cobalt based materials was one of the important discoveries in permanent magnet technology and is still widely used today. These magnets can be magnetized in any direction by simply heating the magnet and cooling them in a magnetic field to give a preferred magnetic direction. Traditionally, Alnico magnets were largely used in PM motors. One advantage of Alnico magnets is that they have a high residual flux density (B_r). They have excellent temperature stability and strong corrosion resistance level. Their working temperatures can go up to 500 degrees. However, they can be demagnetized very easily. In addition, the maximum energy product of these magnets is not very high.

Ferrite magnets, also called ceramic magnets, are one of the cheapest magnets manufactured in industry. They have very high intrinsic coercive force (H_{ci}) and therefore, they are very difficult to demagnetize. They can easily be magnetized in a variety of formats. The raw material is so abundant that it is found in numerous applications. This kind of magnet material has a good resistance to corrosion and can operate at high temperatures up to 300 degrees. These materials are used even today for applications where space and cost are not important requirements.

Rare-earth magnets are strong permanent magnets made from the alloys elements such as Neodymium and Samarium. Discovery of these strong magnets have changed the future of permanent magnet motor technology as well as servomotors and the magnetic field can be increased to 1.5T levels. There are two types of rare-earth magnets available: Neodymium magnets and Samarium cobalt magnets.

The first generation rare earth magnets use Samarium and Cobalt (SmCo). One of the biggest advantages of such magnets is that they provide very high MEP compared to Alnicos and Ferrites. This big improvement in high MEP is made possible by the high coercive force. Nonetheless, they are very brittle and both the raw material cost and the production cost are quite high compared to other types of magnets. The revolution of rare earth magnets accelerated with the discovery of Neodymium Iron-Boron (NdFeB) magnets with even higher MEP in 1982. NdFeB magnets are produced by pressing powders in a magnetic field and their energy products can go up to 420 kJ/m^3. This material is much stronger than SmCo and the cost is much lower simply because they are composed of mostly iron which is much cheaper than cobalt. However, they have to be protected against corrosion and their working temperature is also lower compared to SmCo magnets.

A brief comparison of different magnets used in PM motors is illustrated in Table 4. The rare earth magnets are the most common magnet materials used in PM servomotors and the table clearly shows significant benefits of such magnets. NdFeB magnets have higher flux density levels up to 1.5T and higher MEPs but their working temperature is lower (up to 200 ºC).

Materials	B_r [T]	Hc [kA/m]	BH_{max} [kJ/m^3]	T_C [ºC]	T_{w-max} [ºC]
Alnico	1.2	10	6	500	500
Ferrite	0.43	10	5	300	300
SmCo	Up to 1.1	Up to 820	Up to 240	Up to 820	Up to 350
NdFeB	Up to 1.5	Up to 1033	Up to 422	Up to 380	Up to 200

Table 4. Typical permanent magnet material magnetic properties

NdFeB magnets are more common rare earth magnets than SmCo, cheaper but more brittle. SmCo magnets are widely used in applications in which higher operating temperature and higher corrosion and oxidation resistance are crucial. A basic comparison of the four major types of permanent magnet materials used in motors today is illustrated in Fig. 9.

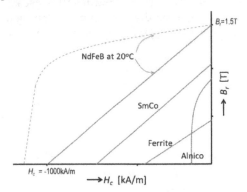

Fig. 9. Flux density versus magnetizing field for the important magnets classes

4. Basic permanent magnet motor design process

Design procedure for any PM servomotor is shown in Fig. 10. This process comprises three main steps: Electromagnetic, structural and thermal designs. Electromagnetic design starts with magnetic circuit modeling and parameter optimization with a given set of design specifications. A series of optimizations such as pole number, loading, current density, dimensional limits etc. have to be performed to find the optimum parameters of the motor before proceeding further. When a design is obtained that meets the technical spec, a quick motor simulation and the influence of parameter variation must be carried out using simulation software such as SPEED (PC-BDC Manual, 2002). A detailed electromagnetic finite element analysis (FEA) either in 2D or 3D is the next step to verify that the design meets the specified torque-speed characteristics and performance. After an electromagnetic design is finalized, structural and thermal analyses (MotorCAD Manual 2004) have to be completed. It should be pointed out that structural analysis is not a necessity at low speed servomotor designs. If the motor does not meet the structural or thermal tests, then the electromagnetic design study should be repeated for a better design. A motor design has to be finalized after a design passes all of the main steps (Aydin et al. 2006).

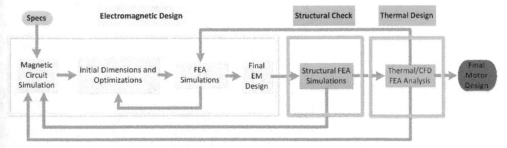

Fig. 10. PM servomotor design process

5. Dynamic model of PM servomotors

Following assumptions are made for the analysis of PM servomotors: The inverter is ideal with no losses; no DC voltage ripple exists in the DC link; the supply current is sinusoidal and no saturation is considered; eddy current and hysteresis loses are negligible; and all motor parameters are constant. Based on these assumptions, permanent magnet servomotor dynamic equations in the synchronous rotating reference frame are written as

$$v_q = r_s i_q + L_q \frac{di_q}{dt} - \omega_e L_d i_d + \omega_e \lambda_f \tag{1}$$

$$v_d = r_s i_d + L_d \frac{di_d}{dt} - \omega_e L_q i_q \tag{2}$$

$$T_m = \frac{3}{2} \frac{P}{2} \left[\lambda_f i_q + \left(L_d - L_q \right) i_d i_q \right] \tag{3}$$

where v_d and v_q are d and q axis voltages, i_d and i_q are d and q-axis currents, r_s is stator resistance, L_d and L_q are d-q axis inductances, ω_e is synchronous speed, λ_f is magnet flux linkage, P is pole number and T_m is the electromagnetic torque of the motor. During constant flux operation, i_d becomes zero and the torque equation becomes

$$T_m = \frac{3}{2} \frac{P}{2} \lambda_f i_q = K_T i_q \tag{4}$$

where K_T is the torque constant of the motor. This equation becomes similar to standard DC motor and therefore provides ease of control. The torque dynamic equation is

$$T_m = T_L + B\omega_m + J \frac{d\omega_m}{dt} \tag{5}$$

equivalent circuit of the steady-state operation of the PM servomotors in d-q reference model is shown in Fig. 11. Independent control of both q-axis and d-axis components of the currents is possible with the vector controlled PM servomotors. Both voltage controlled and current controlled inverters are possible to drive the motor.

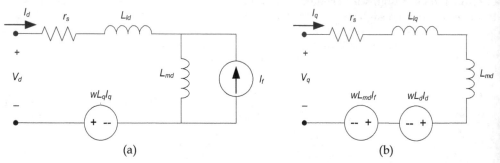

Fig. 11. Equivalent circuit of PM servomotors (a) d-axis equivalent circuit and (b) q-axis equivalent circuit

6. Use of finite element analysis in PM servomotors

The finite element method (FEM) is a numerical method for solving the complex electromagnetic field problems and circuit parameters. It is specifically convenient for problems with non-linear material characteristics where mathematical modeling of the system would be difficult. This method involves dividing the servomotor cross section or volume into smaller areas or volumes. It could be 2D objects in the case of 2D FEM analysis or 3D objects in the case of 3D analysis. The variation of the magnetic potential throughout the motor is expressed by non-linear differential equations in finite element analysis. These differential equations are derived from Maxwell equations and written in terms of vector potential where the important field quantities such as flux, flux direction and flux density can be determined.

The FEM can accurately analyze the magnetic systems which involve permanent magnets of any shape and material. There is no need to calculate the inductances, reluctances and torque values using circuit type analytical methods because these values can simply be extracted from the finite element analysis. Another important advantage of using FEM over analytical approach is the ability to calculate the torque variations or torque components such as cogging torque, ripple torque, pulsating torque and average torque accurately without too much effort.

There are various FEA packages used for motor analysis. FEA packages have 3 main mechanisms which are pre-processor, field solver and post-processor. Model creation, material assignments and boundary condition set-up are all completed in the pre-processor part of the software. Field solver part has 4 main steps to solve the numerical problems. After the pre-process, the software generates the mesh, which is the most important part of getting accurate results. User's experience in generating the mesh has also an important effect on the accuracy of the results. Then, the FEA package computes the magnetic field, performs some analysis such as flux, torque, force and inductance, and checks if the error criteria have been met. If not, it refines the mesh and follows the same steps based on the user's inputs until it reaches the specified error limit. This procedure is shown in Fig. 12. In the post-processor, magnetic field quantities are displayed and some quantities such as force, torque, flux, inductance etc. are all calculated.

7. Torque quality

Permanent magnet servomotors are widely used in many industrial applications for their small size, higher efficiency, noise-free operation, high speed range and better control. This makes quality of their torque an important issue in wide range of applications including servo applications. For example, servomotors used in defense applications, robotics, servo systems, electric vehicles all require smooth torque operation.

One of the most important issues in PM servomotors is the pulsating torque component which is inherent in motor design. If a quality work is not completed during the design stage, this component can lead to mechanical vibrations, acoustic noise, shorter life and drive system problems. In addition, if precautions are not taken, it can lead to serious control issues especially at low speeds. Minimization of the pulsating torque components is of great importance in the design of permanent magnet servomotors.

In general, calculation of torque quality is a demanding task since the torque quality calculation does not only consider the torque density of the motor but also consider the

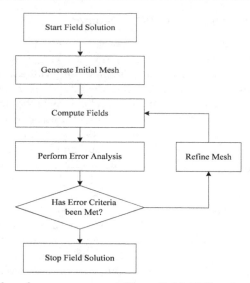

Fig. 12. General procedure for any commercially available FEA software

pulsating torque component. Therefore, a mathematical approach about torque quality should include harmonic analysis of electric drive system rather than a simple sizing of the motor.

Output torque of a PM servomotor has an average torque and pulsating torque components. The pulsating torque consists of cogging torque and ripple torque components. Cogging torque occurs from the magnetic permeance variation of the stator teeth and the slots above the permanent magnets. Presence of cogging torque is a major concern in the design of PM motors simply because it enhance undesirable harmonics to the pulsating torque. Ripple torque, on the other hand, occurs as a result of variations of the field distribution and the stator MMF. At high speed operations, ripple torque is usually filtered out by the inertia of the load or system. However, at low speeds 'torque-ripple' produces noticeable effects on the motor shaft that may not be tolerable in smooth torque and constant speed applications.

Servomotors can also be categorized by the shape of their back EMF waveforms which can take different forms such as sinusoidal and trapezoidal as seen in Fig. 13. Any non-ideal

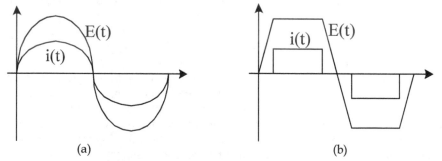

Fig. 13. Current and back EMF waveform options in PM servomotors: (a) sinusoidal back-EMF and current and (b) trapezoidal back-EMF and current

situations such as converter caused disturbed stator current waveform and disturbed back EMF waveform arising from the design cause non-sinusoidal current and airgap flux density waveforms which result in undesired pulsating torque components at the motor output. In other words, if the airgap flux density waveform is disturbed, pulsating torque at the motor shaft becomes inevitable (Jahns & Soong, 1996).

7.1 Electromagnetic torque in PM motors

Definitions of the torque components will be given before getting into the subject any deeper. First, "cogging torque" is defined as the pulsating torque component produced by the variation of the airgap permeance or reluctance of the stator teeth and slots above the magnets as the rotor rotates. In other words, there is no stator excitation involved in cogging torque production of a PM motor. Second, "ripple torque" is the pulsating torque component generated by the stator MMF and rotor MMF. Ripple torque is mainly due to the fluctuations of the field distribution and the stator MMF which depends on the motor structure and the current waveform. This component can take two forms, one of which resulting from the MMF created by the stator windings and the other from MMF created by the rotor magnets. The second form is the torque created by stator MMF and rotor magnetic reluctance variation. In surface mounted PM servomotor, since there exists no rotor reluctance variation, there is no second form of the ripple torque and the ripple torque is mainly created by the first form. The third definition is "pulsating torque" which is defined as the sum of both cogging and ripple torque components (Sebastian et al., 1986 and Ree & Boules, 1989).

In the following analysis, it is assumed that a Y connected three phase unsaturated PM motor used and it has a constant airgap length and symmetrical stator winding. It is also assumed that stator currents contain only odd harmonics and current harmonics of the order of three does not exist. Finally, armature reaction is assumed negligible. For PM motors, at instant t, the instantaneous electromagnetic torque produced by phase A can be written as the interaction of the magnetic field and the phase current circulating in N turns

$$T_a(t) = 2pNi_a(t) \int_{-\pi/2mp}^{\pi/2mp} R_g L_e B(\theta_r, t) d\theta_r \qquad (6)$$

where R_g is airgap radius of the servomotor, L_e is effective stack length, p is pole pairs and m is the number of phases.

The back-EMF of the motor induced in phase A at the time instant t is given by

$$e_a(t) = 2pN \int_{-\pi/2mp}^{\pi/2mp} R_{mean} L_R B(\theta_r, t) \omega_m d\theta_r \qquad (7)$$

where ω_m is the rotor angular speed. The back-EMF in phase a can also be written as

$$e_a(t) = \omega_m 2pN \int_{-\pi/2mp}^{\pi/2mp} R_{mean} L_R B(\theta_r, t) d\theta_r \qquad (8)$$

Substituting (8) into (6), the torque expression becomes

$$T_a(t) = \frac{e_a(t) \cdot i_a(t)}{\omega_m} \qquad (9)$$

For the Y-connected three-phase stator winding, the back-EMF in phase A can be written as the summation of odd harmonics including fundamental component:

$$e_a = E_1 \sin \omega t + E_3 \sin 3\omega t + E_5 \sin 5\omega t + E_7 \sin 7\omega t + \ldots \tag{10}$$

and likewise the current in phase A can be written as

$$i_a = I_1 \sin \omega t + I_5 \sin 5\omega t + I_7 \sin 7\omega t + I_{11} \sin 11\omega t + I_{13} \sin 13\omega t + \ldots \tag{11}$$

where E_n is the nth time harmonic peak value of the back EMF, which is produced by n^{th} space harmonic of the airgap magnetic flux density B_{gn} and I_n is the nth time harmonic peak value of current.

The product of back-EMF and current $e_a i_a$ is composed of an average component and even-order harmonics for all phases. The total instantaneous torque contributed by each motor phase is proportional to the product of back EMF and phase current. In other words, the total instantaneous torque is the sum of the torques produced by phase a, b, and c and given by,

$$T_m(t) = \frac{1}{\omega_m} \left[e_a(t) \, i_a(t) + e_b(t) \, i_b(t) + e_c(t) \, i_c(t) \right] \tag{12}$$

Since the phase shifts between $e_a i_a$ and $e_b i_b$ and between $e_a i_a$ and $e_c i_c$ are $-2\pi/3$ and $2\pi/3$, respectively, the sum ($e_a i_a + e_b i_c + e_c i_c$) will contain an average torque component and harmonics of the order of six. The other harmonics are all eliminated. Thus, the final instantaneous electromagnetic torque equation of a servomotor can be written as

$$T_m(t) = T_0 + \sum_{n=1}^{\infty} T_{6n} \cos n6\omega t \tag{13}$$

where T_0 is the average torque, T_{6n} is harmonic torque components and $n = 1,2,3\ldots$ In the ideal case, if the back-EMF's and the armature currents are sinusoidal, then the electromagnetic torque is constant and no ripple torque exists as illustrated in Fig. 14. The same quantities are plotted for sinusoidal back EMF and square or trapezoidal stator current waveforms and the presence of the pulsating torque component is observed clearly. The resultant plots including torque pulsations for all cases are shown in Fig. 14.

7.2 Cogging torque
7.2.1 Cogging torque theory
Existence of cogging torque is always a cause of concern in the design of PM servomotors. It in fact demonstrates the quality of a servomotor. This torque component is often desired that the motor produces a smooth torque in a wide speed range. Cogging torque adds unwanted harmonic components to both torque output and the torque-angle curve, which results in torque pulsation. This produces vibration and noise, both of which may be amplified in variable speed drive when the torque frequency coincides with a mechanical resonant frequency of the stator and rotor. In addition, if rotor positioning is required at very low speeds, the elimination of cogging torque component becomes even more crucial and must be eliminated completely during the design stage (Li & Slemon, 1988).

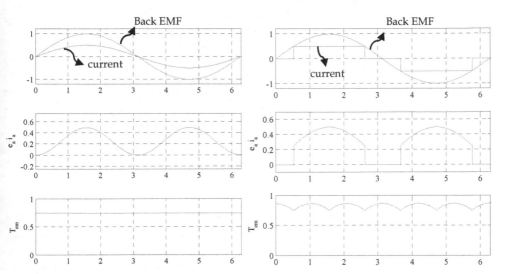

Fig. 14. Torque output for current and back EMF waveforms as a function of electrical cycle

Cogging torque is sensitive to varying rotor position and can be expressed by

$$T_{cog}(\theta_r) = -\frac{1}{2}\phi_g^2 \frac{dR_g}{d\theta_r} \tag{14}$$

where ϕ_g is the airgap flux, R is the reluctance of the airgap and θ_r is the rotor position. Cogging torque increases due to the increased airgap flux as the magnet strength is increased. Nevertheless, the cogging torque results from the non-uniform flux density in the airgap. As the stator teeth become saturated, the flux begins to distribute evenly in the motor airgap and the cogging torque decreases.

In addition, if there is no airgap reluctance variation as in slotless motors, no cogging component occurs. For a slotted stator configuration, the airgap permeance or reluctance is non-uniform because of the shape of the stator, saturation of the lamination material, slot openings and the space between the rotor magnets. This non-uniform reluctance or magnetic flux path causes the airgap flux density to vary with rotor position. This results in cogging torque, and generates vibration.

7.2.2 Minimization of cogging torque component

Cogging torque minimization is a significant concern during the design of brushless PM servomotors, and it is one of the main sources of torque and speed fluctuations especially at low speeds and load with low inertias. A variety of techniques are available for reducing the cogging torque of conventional PM servomotors, such as skewing the slots, shaping or skewing the magnets, displacing or shifting magnets, employing dummy slots or teeth, optimizing the magnet pole-arc, employing a fractional number of slots per pole, and imparting a sinusoidal self-shielding magnetization distribution. A summary of these methods are displayed in Fig. 15 (Bianchi & Bolognani, 2002).

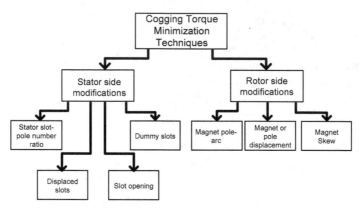

Fig. 15. Summary of cogging torque minimization techniques for PM servomotors

Minimization of cogging torque in PM servomotors can be accomplished by modifications either from stator side or from rotor side. Choosing the appropriate "ratio of stator slot number to rotor pole number" combination is one of the common ways to minimize cogging torque component. This is a design based choice and is the most common method to minimize the unwanted cogging torque components in PM servomotors. Utilizing dummy slots in stator teeth increases the frequency of cogging and reduces its amplitude. Similarly, "displaced slots and slot openings" is a different method to minimize cogging component. In integral slot servomotors (q=1slots/pole/phase), each rotor magnet has the same position relative to the stator slots resulting in cogging torque components which are all in phase, leading to a high resultant cogging torque. Nevertheless, in fractional slot servomotors, where q≠1slots/pole/phase rotor magnets have different positions relative to the stator slots generating cogging torque components which are out of phase with each other. The resultant cogging torque is, thus, reduced since some of the cogging components are partially cancelled out. Even uncommon combinations such as 33, 39 or 45 slots are employed for certain applications to obtain small cogging torque components even though it generates an unbalanced servomotor.

Rotor side cogging torque minimization techniques are more cost effective compared to stator side methods and classified into three different categories: variable or constant magnet pole-arc to pole-pitch ratio, pole displacement and magnet skew. Techniques applied to rotor structure are simpler and less costly than stator side techniques. One of the most effective techniques used in servomotors is to employ an appropriate magnet pole-arc to pole-pitch ratio. Reducing the magnet pole-arc to pole-pitch ratio reduces the magnet leakage flux, but it also reduces the magnet flux, and, consequently, the average torque. Another method of reducing the cogging torque is to employ variable magnet pole-arcs for adjacent magnets such that the phase difference between the associated cogging torques results in a smaller net cogging.

7.2.3 Predicting cogging torque using FEA

Finite element analysis (FEA) can correctly examine the PM servomotors. The motor designers do not need to go through cumbersome circuit type analytical methods because important parameters such as flux, inductance, force and torque can simply and accurately be extracted from the finite element analysis. Even cogging torque component can precisely be calculated using modern FEA software.

Flux 2D software package by Cedrat Co., which is one of the frequently used FEA software in academia and industry, is used in the analyses of the PM servomotor given in Fig. 16 - Fig. 18 (Flux 2D and 3D Tutorial 2002). Cogging torque is obtained using no-load simulations. Rotor structure is rotated for one slot pitch and torque values are calculated using the Flux 2D. Fig. 16 shows both no-load flux density distribution of a 24 slot-8 pole PM servomotor as well as its cogging torque variation over one slot-pitch. Fig. 17 displays the rotor a disc type PM servomotor, its FEA predicted and experimentally verified cogging torque variation. The results show that FEA work well for the cogging torque predictions.

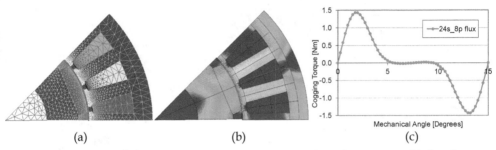

(a)	(b)	(c)

Fig. 16. 2D-FE Model of 24 slots with 8 poles servomotor (a) mesh structure, (b) flux density distribution and (c) cogging torque variation

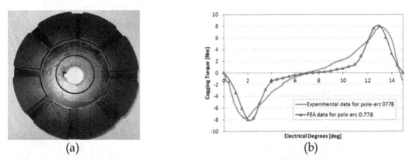

(a)	(b)

Fig. 17. Rotor structure of a disc type PM servomotor (a), prediction of cogging torque with FEA and experimental data (b)

7.3 Torque ripple

Torque ripple is another important undesired torque element in PM servomotors. It occurs as a result of fluctuations of the field distribution and the stator MMF. In other words, torque ripple depends on the MMF distribution and its harmonics as well as the magnet flux distribution. At high speeds, torque ripple is usually filtered out by the system inertia. However, at low speeds torque-ripple may produces noticeable effects on motor shaft that may not be tolerable in smooth torque and constant speed servo applications.

Fig. 18 shows an interior permanent magnet (IPM) servomotor geometry, flux lines and flux density distribution at no load operation. If the motor is supplied by a harmonic free excitation, almost no ripple exists at the motor output (Fig. 19). However, if inverter or motor driven harmonics, such as integer slot motors or single segmented rotors with q=1 with no skew, exist in the current excitation, significant torque ripple appears at the motor output and precautions must be taken to lower this component as much as possible. One of common techniques to reduce the torque ripple component and obtain smooth torque output is to use segmented rotor. In order to use this approach, rotor is divided into segments and each piece is rotated with respect to each other to obtain ripple free output. As displayed in Fig. 19, if no segment is used, more than 130% of torque ripple is observed at the motor output. When the rotor is divided into 4 segments, torque ripple is reduced to less than 6% of the average torque which is a reasonable number for most applications.

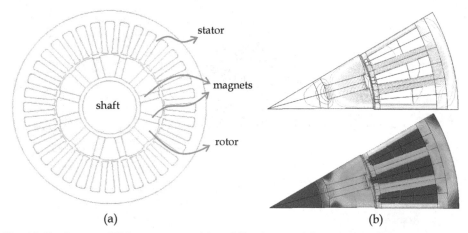

(a)   (b)

Fig. 18. Spoke type IPM servomotor (a) and flux line and flux distribution (b)

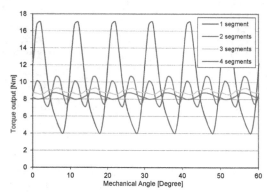

Fig. 19. Torque output variation of the IPM servomotor with low and high torque ripple

8. Control of PM servomotors

Commutation of a brushless pm motor is achieved electronically. Stator winding is energized using an inverter in a sequence. It is crucial to know the position of the rotor so as to know which winding will be energized. This requires precise information of the rotor position using hall sensors or resolvers. When the rotor magnet poles pass, the position information of the sensor is provided to the controller and inverter drives the motor windings in the correct sequence.

Brushless PM servomotors can have both trapezoidal and sinusoidal back EMF waveforms and are excited with either rectangular or sinusoidal currents. A current regulated voltage source inverter is used to drive the servomotors. Power stage of the converter is combined by a rectifier, DC link and an inverter. Current sensors are used in each phase and fed back to the DSP controller. Position information is frequently obtained either by a resolver or an encoder although hall sensors are preferred for trapezoidal brushless servomotors. The simple system set-up with the main blocks of the system is illustrated in Fig. 20.

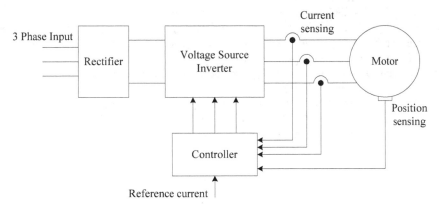

Fig. 20. Permanent magnet servomotor drive

9. Conclusion

Detailed introduction to brushless permanent magnet servomotors used in both industrial and servo applications is provided in this chapter. Motor classification and types, advantages and disadvantages of different PM servomotors and comparison, materials used in motor components are reviewed. Servomotor design process including electromagnetic, structural and thermal steps; softwares used in the analysis, design and optimization of such motors are also enlightened in detail. Torque quality, mathematical output torque equations, cogging and ripple torque components are investigated thoroughly since the torque quality confirms the quality of the servomotor.

10. Acknowledgment

The authors are indebted to CEDRAT Co. for providing the Flux 2D FEA Package and MDS Motor Ltd. for providing some motor pictures and its facilities in preparing this document.

11. References

Flux 2D and 3D Tutorial, Cedrat Co. 2002.

J. D. L. Ree and N. Boules, "Torque production in permanent-magnet synchronous motors," *IEEE Trans. Industry Applications*, vol. 25, no. 1, pp. 107-112, 1989.

J. R. Hendershot and T. J. E. Miller, "Design of Brushless Permanent-Magnet Motors (1995)," Oxford University Press, ISBN 0198593899, UK.

M. Aydin, M. K. Guven, S. Han, T. M. Jahns and W. L. Soong, "Integrated Design Process and Experimental Verification of a 50 kW Interior Permanent Magnet Synchronous Machine", *17th International Conference on Electrical Machines (ICEM 06)*, Crete, Greece, 2006.

Motor-CAD v3.1 software manual, April 2006.

N. Bianchi and S. Bolognani, "Design Techniques for Reducing the Cogging Torque in Surface-Mounted PM Motors", *IEEE Transactions on Industry Applications*, Vol. 38, No. 5, September/October 2002.

SPEED Software, PC-BDC 9.04 User's Manual, February 2010.

T. Sebastian, G. R. Slemon and M. A. Rahman, "Design considerations for variable speed permanent magnet motors", *Proceedings of International Conference on Electrical Machines (ICEM)* 1986, pp.1099-1102.

T.Li, and G. Slemon, "Reduction of cogging torque in PM motors," *IEEE Trans. Magnetics*, vol. 24, no. 6, pp 2901-2903, 1988.

Thomas M. Jahns and Wen L. Soong, "Pulsating Torque Minimization Techniques for Permanent Magnet AC Motor Drives-A Review", *IEEE Transactions on Industry Applications*, pp. Vol. 43, No. 2, April 1996.

3

Estimation of Position and Orientation for Non–Rigid Robots Control Using Motion Capture Techniques

Przemysław Mazurek
Department of Signal Processing and Multimedia Engineering
West Pomeranian University of Technology, Szczecin
Poland

1. Introduction

Robots are well established in science and technique. They are used in different environments and they have different structures. Typical robot movements are rapid and steepy when the movement direction changes occurs. It is not necessary to replicate a biological nature based solution for most tasks, so such movements are acceptable and simpler to obtain. Control algorithms are simpler for such cases, development and settings of such controllers are more straightforward.

Robots are also based on a set of joins and serial or parallel configurations. Different configurations are usefully for selected task and may be not based on biological nature references. Replication of biological nature are not necessary, and for example a wheels that are simple to design have not biological references.

Join based approach of robot design is well established and there are many technical advantages of such structure (Fig. 1). Mechanic of the robots is based mostly on a kind of the skeleton. The endoskeletons design uses a mechanical parts located inside light–weight casing. The exoskeletons design uses a mechanical parts that is casing also. Some robots uses mixed design, where the 'bones' are in endosceleton design and only joints uses exosceleton design. Exosceletons design are used in hostile environments typically.

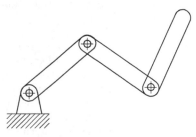

Fig. 1. Rigid actuator

The bones are fixed, so length of the bone or its curvature is not possible to change. Additional actuators for arm extension are used sometimes. Fixed structure of the robot, even if redundant number of degrees of freedom is available, is convenient for analysis and design.

2. Non–rigid robots

Rigid structure is not only–one solution for the robots. The flexible (non–rigid, elastic) robots, complete actuators, and partially flexible actuators are also important for the future robots (Fig. 2). Flexibility of the actuators or overall robot's body is inspired by the biological nature. The giant amount of the species that live in different environments uses flexible bodies or bodies parts with evolutionary success.

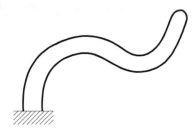

Fig. 2. Non–rigid actuator

Non–rigid robots are active and open research area. Any physical effect related to the flexible movement that is driven by the any factor (continuously or PWM–like) may be applied for intentional movement of robot or actuator. Direct or indirect control of the movement by the electrical signal is desired especially. The Pulse Width Modulation (PWM) control is especially important for simplification of driver.

The pneumatic[Daerden (1999); Daerden & Lefeber (2002); Verrelst (2005)] or hydraulic effects may be used, but the control of them is possible by the electric–to–pneumatic or electric–to–hydraulic conversion devices and is indirect control. The bimetal or memory alloy actuators are controlled by the electricity more directly (exactly there is an electrical energy–to–heat conversion but no additional devices like pumps are necessary).

Progress in the material engineering and market availability of materials, that are sensitive on electricity, gives an ability of application such 'muscles' for skeleton based robots, or even a build an complete body of non–rigid robots.

Non–rigid robots may be characterized by the place of movement. In the rigid robot the movement points are defined by the main gear axis or motor axis. For linear movement the line of the movement is also well defined like for the linear stepping motors for example. Conversion of rotary to linear movement is also used.

The non–rigid robots may be controlled in a hundreds points per muscle. The conventional classification and comparison of actuators, based on the number of degree–of–freedom, is not convenient for such cases. Electrical based control of Electroactive Materials gives building abilities of robots and controlling them in so many of points, giving a new way of robots design. Such robots may change external shape and size (morphing robots).

A few main types of Electroactive Materials are used and developed nowadays.

The simplest are the bimetal strips and coils based on the conversion of electrical energy into heat. The Shape Memory Alloys are also interesting alternatives to bimetal, and the best know is the Nitinol (Nickel titanium). The more advanced actuators like Biometal helix, due significant length changes are also important. The Nitinol was applied in well know Stiquitio hexapod robot legs and derivatives [Conrad & Mills (2004)]. The main drawback of the bimetal and SMA is the speed of the physical changes that is about a few seconds depending on material and design. Heating of such material is controlled by the electricity and could be very rapid, but cooling is depends on the environment of the work.

The Electroactive Polymers (EAPs) are the most promising materials for non–rigid robots design nowadays. The advantages of such materials are applied for rigid robots also, because it is important replacement of the electrical motors. Improved reliability, increased lifetime, reduction of electromagnetic emission are very important for robot design. There are many materials based on the different effects [Bar–Cohen (2004); Besenhard et al. (2001); Capri & Smela (2009); Chanda & Roy (2009); Hu (2007); Kim & Tadokoro (2007); Otake (2010); Wallace et al. (2009)]. The Electronic EAPs uses piezoelectric, electrostatic, electrostrictive and ferroelectric effects nowadays. The Ionic EAPs uses the displacement of ions inside the polymer.

One of the most important task is the measurement of the state of such robot or manipulator. The conventional position and orientation approach is not well fitted, because non–rigid robots are flexible, so huge or infinite number of positions points are possible. Moreover, the estimation of the number of degrees–of–freedom by the simple visual observation of robot movements is not feasible.

The reasonable way is to estimate position and orientation in some points, especially for the end–effector and limited number of selected intermediate points. The overall estimation is possible, using the model based techniques and vision measurements. The vision techniques are well suited for such robots, because they make measurements in hundreds or millions points (pixels in extended cases).

The open–control loop, without knowledge about achieved state, is applicable for very specific cases only, for non–rigid robots. The flexibility of the non–robots have important disadvantage – the forces from manipulated objects and forces from environments influent on the achieved state. Such forces change state and in the worst case all points of the non–rigid robots may differ between the expected position and real one. This is one of the reasons why the closed–control loop for rigid robots and the state estimation are necessary. Vision based technique for rigid robots (visual servoing [Agin (1979); Chaumette (1998); Chaumette & Hutchinson (2008); Corke & Hutchinson (2001); Fung & Chen (2010); Malis at al. (1999); Marchand at al. (2005); Sanderson & Weiss (1983)]) are used from many years, and it is very promising technique for non–rigid robots also.

3. Visual systems for non–rigid robots

Different video tracking schemes for non–rigid robots and actuators are possible, and the selected are presented shortly.

3.1 Conventional motion capture system

Conventional motion capture system (multiple camera vision system [Aghajan & Cavallaro (2009)]) uses a set of cameras located around robot (Fig. 3). Video tracking gives abilities of the robot state estimation what is necessary to control. Such system is very simple for implementation in comparison to other presented tracking schemes. The market availability of such systems for large working area (known as a volumen) like a cubic area with a few meters distance in every direction is important for large scale systems. There are also available systems for small working area about half meter in every direction.

Typical motion capture system uses markers for estimation of the state of human or some objects. The measurements are contact less so significant integration or embedding into robot surface is not necessary. The weight of the robot is preserved. Motion capture system may be used for measurements a very large number of points located on the robot surface. Single or a few cameras are sufficient for estimation of the robot state in most cases.

There are also drawbacks related to the vision techniques. Occlusion reduces a possibility of the state estimation, and the multiple cameras are necessary for reduction of such effects, but

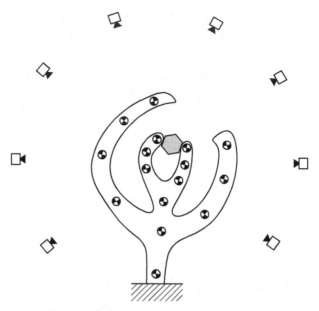

Fig. 3. Motion capture of the non–rigid robot

elimination of the occlusion is not possible for general scenario. Occlusions may occurs due to self occlusions of the robot by the own parts like arms, or may occurs if the environment or operated objects are close to robot.

Internal parts of the robot and related states are hard to estimate, if the cameras are placed around robot. The estimation of the state is based on the outer surfaces, and the estimation of the inner parts of the robot is very difficult task. Such situation occurs for example in estimation task for the propulsion part of the underwater robot that is based on the biological nature species like squids. This is specific design problem, but should be identified on early development stage.

Illumination of the working area influent on the estimation result. The constant environment conditions are recommended. The variable conditions, like bright light sources, may disturb image acquisition by overexposure. Constant light conditions are especially important for the retroreflective markers. Light emitting markers are more robust for variable illumination conditions. Overexposure and underexposure conditions needs expensive HDR (High Dynamic Range) cameras.

High speed cameras are available today (fps > 100 or 1000) but the latency is also very important factor for smooth control of robot, so the image processing part should be integrated into sensors (intelligent cameras are recommended). Most professional motion capture systems uses image processing of acquire image inside camera for bandwidth reduction between camera and computer. Marker detection algorithms are processed in hardware, for reduction processing costs on computer, moreover. High speed cameras reduce distance between position of the marker on two following frames, what gives ability of application simple marker tracking algorithms and assignment algorithms. Gate based approach and nearest neighborhood algorithms for assignment are an example of the simple but an effective algorithms. Assignment is necessary for the tracks of markers maintenance. Assignment is simpler to do if the markers are more unique complex patterns. Position, scale and rotation invariant markers may consist information about unique number of marker. Larger markers due to additional information about number are less usefully due to size, but the color coded information about number is interesting alternative.

Image processing and state estimation algorithms should be the low–latency and real–time. Fixed processing time or variable with known maximal response time is necessary. Detection, tracking and assignment algorithms should be carefully selected.

The commercial motion systems are mostly closed design, without possibility of the algorithm replacement. There are no available free systems, contemporary.

The conventional systems based on the multiple cameras is not unique, and the similar idea based on the video based estimation is possible for other configurations. Most advantages and disadvantages are preserved in other configurations. Some of them are interesting for new robot design.

3.2 Robot equipped with the vision systems

Some, especially mobile robots, uses own vision system for navigation purposes and objects manipulations. The availability of the own vision systems is important for inspection robots working in an hostile environments, especially for space probes, or planet exploration robot.

Vision systems are used also for remote examination of the current state of the robot in case of the significant motion error. Blocked wheels, failed arms or legs due to unexpected environment case or own failure are possible to detect using vision system used for navigation purposes or objects manipulation. This is typical procedure in space robots nowadays. Vision sensors placed on the flexible arms helps in such situation (Fig. 4), gives ability of failure source inspection, finding solutions and may save (extend live) of a multi million dollar robot.

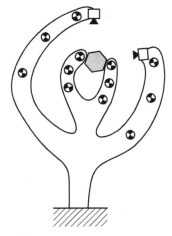

Fig. 4. Non–rigid robot equipped with the vision systems

The conventional sensors may fail and the availability of the vision system gives ability of redundancy also. Proper design uses cameras for the navigation and manipulation task, and many other sensors for movement control. Secondary task of vision system are measurements of state for motion control in case of failure of primary motion control (measurement subsystem).

The non–rigid robots gives interesting ability of application vision system using own multiple cameras. Such robot may change own state and additionally camera position and orientation, creating a different camera configurations on demand. The concept of such robot is similar to amoeba, that have large ability of the shape modification.

The cameras are integrated into robot's flexible body. Range of the work is unlimited and not limited to the unique area (volumen). Different camera configurations may be proposed in real–time and tested for optimal object manipulation or movement. The most challenging task is the multiple or single camera calibration [Daniilidis & Eklundh (2008); Lei et al. (2005); Mazurek (2010; 2009; 2007)]. The estimation of external parameters is especially important for such robots.

3.3 Video sensors on robot's surface
This is specific version of the previous case and inverse motion capture configuration. The cameras are placed on the robot and the fixed set of markers is observed by them. The robot environment is used for the robot's state estimation (Fig.5).

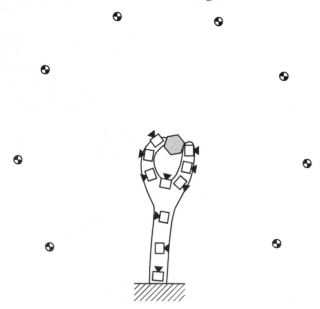

Fig. 5. Cameras placed on the robot surface

It is possible to use cameras for navigation and manipulation purposes and for estimation own state also.
One of the most important factor is the power consumption for such case. The motion capture configuration using passive markers on robot does not need additional power for robot. Inverse motion capture system needs a power supply for camera and acquisition devices. Image processing for inverse case by the external computer is important technique for reduction of the needs of the additional electrical power. The weight of the robot is reduced if the computational part is outside of robot.

3.4 Cooperative robot swarm with multiple cameras
Another possibility when multiple robots (rigid on non–rigid) are equipped with cameras for navigation, manipulation and self measurements. Swarm members are separated robots from the physical point–of–view, but from the logical point–of–view it is a single robot if the cooperation between members of swarm is very close. The self measurement task (Estimation of own parameters) is very interesting, because the state of the particular member of the

swarm is obtained from neighborhoods members. Favorable members are inside swarm due to availability of multiple views (multiple independent observations) from neighborhood members. The outer members are partially observed only. The multiple swarm members may cooperate in many ways.

4. Vision based estimation of position and orientation

The images acquired by the camera set, gives information about 3D world using multiple 2D views. Relation between image objects or additional knowledge about object may be used for estimation of position objects and camera. Without additional knowledge a relative, spatial relations are obtained.

4.1 Features and model based approaches

The vision techniques use feature points or model fitting approaches. Both of them are important for establishing relations between real and virtual (computer modeled) world. In the case of motion capture systems the markers (feature points) are placed on robot or deformable model of robot is used (model fitting).

Feature points are existing features of surrounding object in environment (e.g. corners, edges) or intentionally added (e.g. ball shaped markers, or painted chessboard patterns). Estimation of the position (for point like features) and optionally orientation (for edges or patterns) gives ability of estimation of camera position relative to object.

The model fitting approach is based on the 3D model of robot. The camera measurements are related to the estimation of the pixel assignment to the background or robot body. The aim of the fitting is to find the configuration of the model, that gives image for single camera system or images if multiple camera system are used. The corresponding real and virtual (rendered) images are fitted if the configuration of real robot and its model are identical.

4.2 Correspondence by the calibration object

The simplest technique that is used for establishing relations between virtual and real camera is based on the calibration object. This techniques uses physical object with known physical dimension (M) and mathematical model of this object (V). The bridge between the real and virtual world is the calibration object and its model (Fig. 6).

Assuming, that the worlds coordinates (O, X, Y, Z) are defined if fixed relation in virtual and real calibration object, the full correspondence may between objects, projections and cameras is possible (degenerative cases are not considered here). It means, that all particular positions and orientation have exact values. The projections are the images of the markers from the cameras. Acquired image from the real camera is processed for the marker's positions estimation with subpixel accuracy (e.g. center of mass algorithm may be used). The projection of the virtual markers (V) on the virtual camera projection plane is possible using the computer graphics formulas using high, usually floating point accuracy.

During the estimation process of the external parameters the camera, the correspondence is obtained with some error. Markers projections are not identical and cameras parameters are not equal, especially in beginning steps. The error (Fig. 7) between projections (m, v) of markers (M, V) is possible to calculate. Comparison of the 2D positions on projection planes using l_2 value is used typically (Euclidean distance). Iterative calculations with subject of the minimization of this error are used for establish reliable correspondence.

The accumulative l_2 error is computed using the following formula:

$$l_2 = \sqrt{\sum_i d_i^2} = \sqrt{\sum_i (m_i - v_i)^2} \tag{1}$$

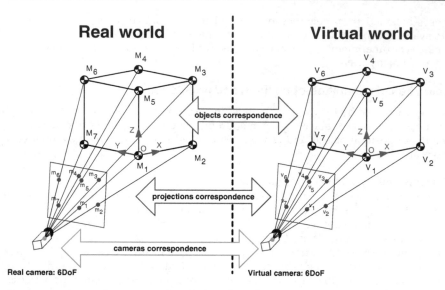

Fig. 6. Correspondences between real and virtual world using 3D calibration object

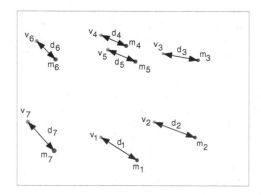

Fig. 7. Comparison of the markers' positions (real and virtual) and local distance errors

Minimization process of l_2 value by the movement and rotation of virtual camera is possible using gradient and non–gradient search algorithms. The difference between position of markers' projections d_i are reduced to zero only in ideal case. The estimated position of the real markers is obtained with some accuracy due to acquisition errors (image blur, finite resolution of the imaging senor, camera noises, design of the imaging sensor, and estimation algorithm for the position).

Estimation of the 3D position and orientation using 2D images is possible using the projective geometry, but the application of Euclidean geometry is also possible. Euclidean geometry is a subset of the projective geometry and preserves angles. Using the long focal length camera, for high ratio of the camera distance to the robot work area is possible.

The Euclidean approach is simpler and cheaper for some cases, especially if the robot is very small. Required large distance between camera and object in real scenarios is main drawback (Fig. 8). Estimation of the 3D position is necessary using a few cameras.

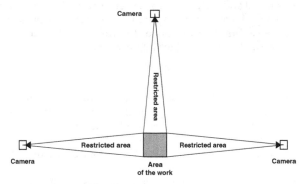

Fig. 8. Example configuration of three cameras for Euclidean geometry based 3D estimation system

The restricted areas and large distance between camera and area of the work requirements are drawback of the Euclidean projections. The perspective projection (Fig. 9) is more applicable for a general case of cameras and different work area configurations.

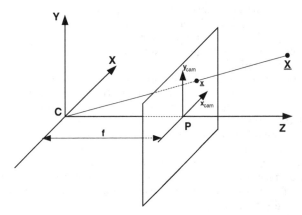

Fig. 9. Perspective projection

Perspective projection uses camera with focal point at position C and projection plane located at distance f, that is focal length. Depending on distance between focal point and point X in the 3D space, the projection x is in different position on projection plane. The projection formulas of point X on projection plane of camera located in arbitrary position in 3D space are available in many computer graphics books [Hartley & Zisserman (2003); Heyden & Pollefeys (2004)].

The perspective projection adds a very important factor – the scale for non–point objects, especially markers. The scale and distance from camera estimation is possible using single camera, depending on the assumed marker estimation technique. Commercial motion capture systems uses very small markers and wide angle cameras (short focal length). The distance is not well measured, especially for variable light conditions for such configuration. The scale of the larger marker may differ in some direction, so for example the ellipse is observed instead the circle. It gives an ability estimation of full 6 DoF (Degree–of–Freedom) for every large marker.

Correspondence between real and virtual world is used during the calibration of the cameras. Calibrated cameras are used in marker systems or in model fitting approach. The model fitting approach is similar to the calibration process but the instead calibrated object there are two sets of calibrated cameras (real and virtual) and deformable model of the real robot.

5. Markers

Vision tracking techniques for robots are based on the numerous approaches: marker based, object features, or even on complete synthesis of the expected object. All of the them are interesting and the selection is application depended. The most valuable techniques for the controlled environment scenarios are marker based. The uncontrolled environments exist if the unexpected situations may occurs, related to the object occlusions, different lighting conditions, etc.

The marker based techniques are very interesting, because different markers designs are possible. The light emitting markers are especially useful for poor lighting conditions. They need additional power connections (wires) for the bulbs or LEDs. The retroreflective markers reflect surrounding light and no additional power connection are necessary for them. The retroreflective markers are interesting for small size and small power robots, especially.

Controlled environment of the robot's work area gives an ability of the correct light setup for maximization performance of retroreflective markers. Markers may support angular estimation (3DoF) depending on own shape. The simplest matte ball markers are orientation less so only a 3D position (3DoF) is obtained by the triangulation using two or more cameras. The carefully selected set of such markers located at close distance gives ability of estimation of orientation. The larger markers with additional orientation features may support estimation of orientation.

In this paper, the four–sector circle with the boundary ring is used as marker (Fig. 10). Such marker gives an ability of orientation estimation with 180 degree accuracy, position and distance. Complete set of DoF (six of them) is possible to estimate. The estimation of all parameters is limited by the optical visibility of the markers. A low angle case between camera and marker plane are hard to process. This is the reason, why a ball shape markers are preferred, because they have superior visibility. Large markers support estimation of own parameters even for partial occlusions but it is not considered in following tests.

The marker uses boundary ring for improving separation between background and marker, what is important for the scale estimation, because estimation process should be related to the marker, not to the background.

6. Estimation of the position, distance, and rotation of the markers

Estimation of the marker is possible using numerous image processing techniques. The feature based techniques or image synthesis, using the model fitting, are possible also. The feature techniques are based on the corners detection. More advanced techniques uses corners detection for further starting point of the line detection. The estimation quality depends on the marker shape and the number of pixels used for the estimation. The larger number of pixels and larger distant between used pixels are important due to noises and possible occlusions, especially.

Proposed technique use a few techniques for the optimization algorithm. Single technique is not feasible for applications due to computation cost and poor results as it is shown by

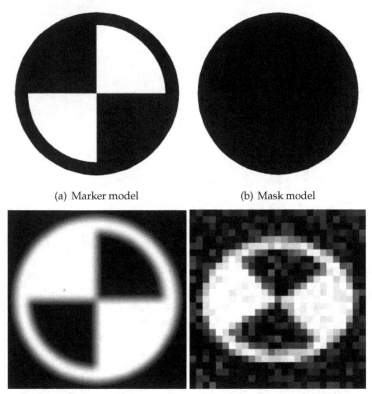

(a) Marker model   (b) Mask model

(c) Filtered (blurred) marker before (d) Example of noised image of
downsampling   marker at low resolution

Fig. 10. Model of marker and noised measurements

the numerical tests. The dedicated renderer of the marker and mask at different positions, scale (distance), 3D rotations, and contrast is used. The contrast fitting is important due to variable light conditions. The white and black points are defined by the two coordinate pairs (Fig. 11). The black (Xb, Yb) and white point (Xw, Yw) define simplest contrast, brightness, and saturation parameters of image transformation.

The first optimization phase is quite simple and the exhaustive search is used for a priori defined spatial and angular resolutions. Positions are tested using subpixel resolutions, 10 times higher resolution in both direction, and rotations using 20 deg. angle resolutions. The scale is not tested, because different scales of markers have common central part. Contrast is also not tested and fixed. The advantages of this phase are the fixed computation cost and possibilities in parallel processing.

The best position obtained from first phase due to obtained l_2 value is tested using optimization in second phase. The selection is driven by the threshold value for l_2 value. Second phase is started in parallel for obtained positions with enough low value of l_2. Second phase is based on the gradient and non–gradient approaches. The constrained optimization is applied in all optimization phases.

During second phase gradient search algorithm is used and after the optimization is stopped (due to achieving error small changes, or after selected number of iterations) the non–gradient

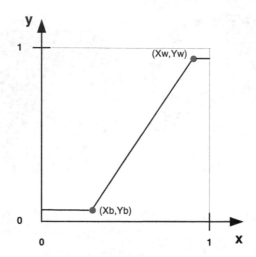

Fig. 11. Contrast curve, with white (w) and black (b) points

algorithm is started. This process is iterated ten times. Such technique gives abilities of exit from local minimal, that is achieved by the gradient search. The non–gradient algorithm when is used alone, supports exit from local minimum (but the convergence is usually very slow).

The gradient algorithm is the minimization procedure from the Matlab Optimization Toolbox (fmincon). The non–gradient algorithm is evolutionary algorithm [Back (T. et al.;T); Michalewicz (1996)], based on mutation. The single parent and child are used at one time evolutionary step. Mutation operator changes relative values of estimated parameters in specific range [Spears (2000)].

The non–gradient phase (Fig.13) uses 1000 iterations and during the single iterations modification of the position (2 DoF), scale (2 DoF), rotation (1 DoF), and contrast (4 DoF) are driven by the uniform random noise generator. The probabilities of mutation of parameter is set to the 0.3. More then one parameter may change during the single iteration. Multiple parameters modified during one iterations reduce influence of local minimum.

The number of iterations and number of repetitions is selected after a lot of tests. The convergence to acceptable level of l_2 is obtained in most cases, but as it is shown later the better results are obtained, if more such optimization processes are started. In parallel processing devices reduce processing time.

7. Performance of proposed estimation technique

Monte Carlo approach Fishman (2000) is used for performance analysis. Application of the Monte Carlo method gives an abilities of testing complex system. The 600 tests are applied using pseudo random number generator for parameters setting. Every test uses 20 iterations (gradient and non–gradient). The Gaussian additive noise is applied to the image (0.2 standard deviation). Values that are not fitted into $(0 - 1)$ range are processed by the contrast curve and saturated according this curve.

The l_2 error is minimized to low values (Fig. 14) what is a numerical, Monte Carlo test based proof of algorithm. Achieving a zero value of l_2 is a very low probable, dependent on the noise level and contrast curve. It means that l_2 error is interesting quality of fitness, that is available during optimization process (due to known model) but not necessary a reliable one.

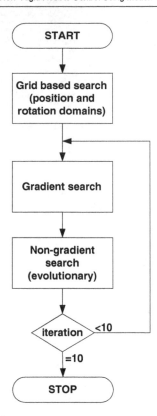

Fig. 12. Optimization scheme

Very interesting are box–plot statistic (Fig. 15), especially the depicted median value of the l_2 error. It is shown, that first step (gradient based) reduces error to low–level, but second step (non–gradient) reduces error to much lower level (more then 2 times). The reduction does not occur significantly by the next repetition of the gradient and non–gradient search. It is very important for practical applications. The gradient algorithm fails in local minima and the solution is possible using the non–gradient algorithm. Applications of the non–gradient algorithm only is not shown in this chapter, and the computation cost is very large (the computation are very slow, and are omitted).

The position error and following errors are calculated using Euclidean distance formula also. All of them are possible to obtain using the synthetic test using Monte Carlo technique and gives an ability of the algorithm test and configuration.

First gradient step does not gives good results (Fig. 16). The mean value of the position error is about half of pixel. The next step (non–gradient based) reduces mean error to values about 0.2. It is important quality improvement. The next steps reduces error, but not significantly. After all 20 steps the mean value is reduced, and histogram is little compressed into left direction, but the computation cost is quite high.

Only z–axis is considered in shown results, that is related to the rotation of the maker around own axis. Rotation errors are reduced significantly (Fig. 17) to the about 0.6 degree (mean value). The reduction occurs after 20 iterations but is not so large.

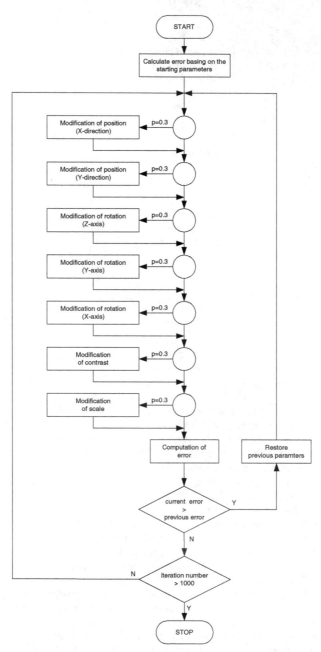

Fig. 13. Evolutionary optimization

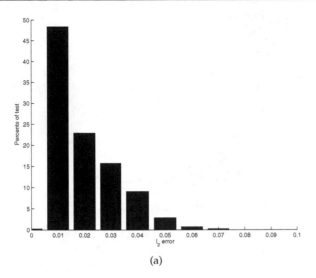

(a)

Fig. 14. l_2 error between image and marker's image model

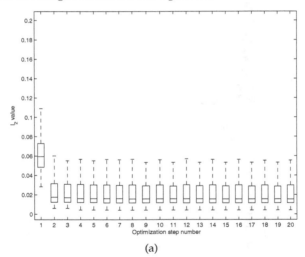

(a)

Fig. 15. Boxplot statistic for optimization step number – gradient based (odd step numbers), non–gradient based (even step numbers)

Known marker size give abilities of distance estimation using single camera. The 24 pixels of diameter correspond to the scale value 0.03. Diameter has variable diameter 24 − 48 pixels in test. The absolute scale error (Fig. 18) is large – 2'nd and 3'rd column preserved about 1/3 cases for errors about 20% of diameter.

Correlation between l_2 value and particular error is very interesting. The following values are obtained: $R = 0.24$ for position, $R = 0.12$ for rotation, and $R = 0.30$ for scale. All tested cases are depicted in Fig. 19 .

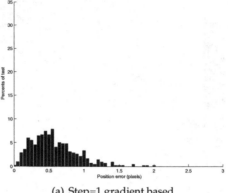

(a) Step=1 gradient based

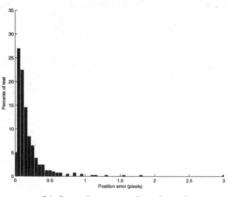

(b) Step=2 non–gradient based

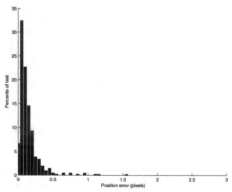

(c) Minimal value after 20 steps (10 iteration of gradient and non–gradient interleaved algorithms)

Fig. 16. Position error

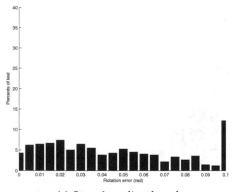

(a) Step=1 gradient based

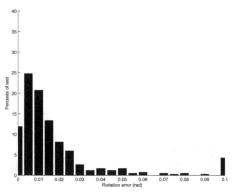

(b) Step=2 non–gradient based

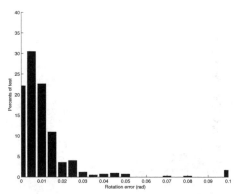

(c) Minimal value after 20 steps (10 iteration of gradient and non–gradient interleaved algorithms)

Fig. 17. Rotation error (z–axis)

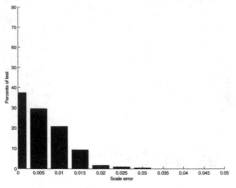

(a) Step=1 gradient based

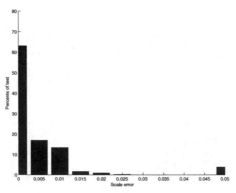

(b) Step=2 non–gradient based

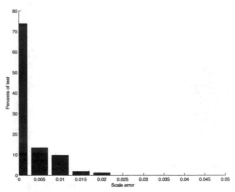

(c) Minimal value after 20 steps (10 iteration of gradient and non–gradient interleaved algorithms)

Fig. 18. Scale (distance) error

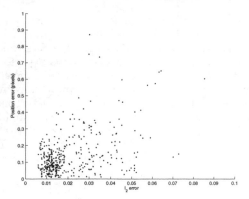

(a) Correlation between image l_2 and position errors

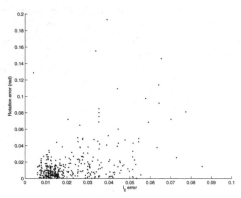

(b) Correlation between image l_2 and rotation errors

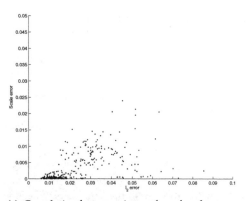

(c) Correlation between image l_2 and scale errors

Fig. 19. Correlation between image l_2 and position, rotation, scale errors

Come discussion is necessary, because correlation coefficients are quite low. there is not direct relation between both errors, like $R = 1/4$, but correlation is high. Most cases create concentration cloud around expected minimal values of both errors. It is well visible for position errors, and if very small l_2 error is measured it is expected that position error is very small also. It looks that position error gives quite large number of pixels used during estimation and the number of pixels influence on the optimization results. The number of pixels for rotation influenced by the rotation probably is lower. This hypothesis should be considered in further works. Improvement of correlation gives abilities of error estimation. This is very important for the tracking algorithms (e.g. Kalman filter) and validation of the measurements. The scale is quite specific, because difference between different scales is defined by the marker's ring. Applications of larger chessboard marker should give better results, but the large markers are not feasible to use.

8. Conclusion

Non–rigid robots are important design for the future robots and different vision based techniques should be applied for the state estimation. Considered technique, based on the larger marker, is promising for state estimation measurements. Estimation of all parameters is possible but the position, rotation, and scale are considered only. The low values of l_2 error corresponds to the low values of the position, rotation and scale errors, and it is a useful estimator of the fitness, but not ideal. The most interesting result is the search scheme, based on the subpixel testing (0.1 pixel accuracy), gradient search and non–gradient search. The next repetition of gradient and non–gradient algorithm does not reduce error so much. Estimation of the parameters (meta level optimization) of non–gradient algorithm is interesting. The 1000 steps are used and reduction of the number of steps is important for the real–time processing. Validation of the proposed algorithm is important for the further optimization and in parallel processing. At this moment, processing time is quite long using Matlab. Optimization of the algorithm and code is necessary together. The visual servoing applications need fast, low latency and computation cost effective solutions. Application of the GPGPU or FPGA are promising computation devices for considered algorithm.

9. Acknowledgments

This work is supported by the UE EFRR ZPORR project Z/2.32/I/1.3.1/267/05 "Szczecin University of Technology – Research and Education Center of Modern Multimedia Technologies" (Poland).

10. References

Aghajan, H. & Cavallaro, A. (2009). *Multi–Camera Networks. Principles and Applications*, Academic Press, ISBN 9780123746337.

Agin, G.J. (1979). *Real Time Control of a Robot with a Mobile Camera* Technical Note 179, SRI International.

Back, T., Fogel D.B., Michalewicz Z. (2000). *Evolutionary Computation 1. Basic Algorithms and Operators*, Institute of Physics Publishing, ISBN 0750306645.

Back, T., Fogel D.B., Michalewicz Z. (2000). *Evolutionary Computation 2. Advanced Algorithms and Operators*, Institute of Physics Publishing, ISBN 0750306653.

Bar–Cohen, Y. (Ed.) (2004). *Electroactive Polymer (EAP) Actuators as Artificial Mulscles. Reality, Potential, and Challenges*, SPIE Press, ISBN 0819452971.

Besenhard, J.O., Gamsjäger, H., Stelzer, F., Sitte, W. (2001). *Electroactive Materials*, Springer–Verlag KG, ISBN 978-3-211-83655-2.

Capri, F., & Smela, E. (Eds.) (2009). *Biomedical Applications of Electroactive Polymer Actuators*, John Wiley & Sons, ISBN 978-0-470-77305-5.

Chaumette, F. (2008). Potential problems of stability and convergence in image–based and position–based visual servoing, In: *The confluence of vision and control, volume 237 of Lecture Notes in Control and Information Sciences*, Kriegman, D., Hager, G., Morse, S., (Eds.), 66–78, Springer–Verlag, ISBN 1852330252, New York.

Chaumette, F. & Hutchinson, S. (2008). Visual Servoing and Visual Tracking, In: *Handbook of Robotics*, Siciliano, B. & Khatib, O., (Eds.), 563–584, Springer, ISBN 978-3-540-23957-4.

Chanda, M. & Roy, S.K. (2009). *Industrial Polymers, Specialty Polymers, and Their Applications*, CRC Press, ISBN 9781420080582.

Conrad, J.M., & Mills, J.W. (1997). *Stiquito(TM): Advanced Experiments with a Simple and Inexpensive Robot*, IEEE Computer Society Press, ISBN 0818674083.

Corke, P. & Hutchinson, S.A. (2001). A new partitioned approach to image–based visual servo control. *IEEE Transactions on Robotics and Automation*, Vol. 17, No. 1, Aug – 2001, 507–515, ISSN 1042-296X.

Daerden, F. (1999). *Conception and realization of pleated pneumatic artificial muscles and their use as compliant actuation elements* PhD Dissertation, Vrije Universiteit Brussel, http://lucy.vub.ac.be/publications/Daerden_PhD.pdf.

Daerden, F. & Lefeber, D. (2002). Pneumatic artificial muscles: actuators for robotics and automation. *European Journal of Mechanical and Environmental Engineering*, Vol. 47, No. 1, 2002, 10–21, ISSN 1371-6980, http://lucy.vub.ac.be/publications/Daerden_Lefeber_EJMEE.pdf.

Daniilidis, K. & Eklundh, J.–O. (2008). 3–D Vision and Recognition, In: *Handbook of Robotics*, Siciliano, B. & Khatib, O., (Eds.), 543–562, Springer, ISBN 978-3-540-23957-4.

Fishman, G.S. (2000). *Monte–Carlo. Concepts, Algorithms, and Applications*, Springer, ISBN 038794527X.

Fung, R.–F. & Chen, K.–Y. (2010). Vision–Based Control of the Mechatronic System, In: *Visual Servoing*, Fung, R.–F., (Ed.), 95–120, Intech, ISBN 978-953-307-095-7.

Hartley, R. & Zisserman, A. (2003). *Multiple View Geometry in Computer Vision*, Cambridge University Press, ISBN 0521540518.

Heyden, A. and & Pollefeys, M. (2004). Multiple View Geometry, In: *Emerging Topics in Computer Vision*, Medioni G., & Kang, S.B, (Eds.), 45–108, Prentice Hall, ISBN 9780131013667.

Hu, J. (2007). *Shape memory polymers and textiles*, CRC Press, ISBN 1845690478.

Kim, K.J. & Tadokoro, S. (Eds.) (2007). *Electroactive Polymers for Robotic Applications. Artificial Muscles and Sensors*, Springer, ISBN 184628371X.

Lei, B.J., Hendriks, E.A., Katsaggelos, A.K. (2005). Camera Calibration for 3D Reconstruction and View Transformation, In: *3D Modeling and Animation: Synthesis and Analysis Techniques for the Human Body*, Sarris, N. & Strintzis, M.G., (Eds.), 70–129, IRM Press, ISBN 1591402999.

Malis, E., Chaumette, F., Boudet, S. (1999). 2.5 D visual servoing. *IEEE Transactions on Robotics and Automation*, Vol. 15, No. 2, 1999, 238–250, ISSN 1042-296X.

Marchand, E., Spindler, F., Chaumette, F. (2005). 3–D Vision and Recognition, In: *Handbook of Robotics*, Siciliano, B. & Khatib, O., (Eds.), 543–562, Springer, ISBN 978-3-540-23957-4.

Mazurek, P. (2010). Mobile system for estimation of the internal parameters of distributed cameras. *Measurement Automation and Monitoring* Vol.56, No.11, 1356–1358, ISSN 0032-4110.

Mazurek, P. (2009). Estimation of state–space spatial component for cuboid Track–Before–Detect motion capture systems. *Lecture Notes in Computer Science*

Vol. 5337 (Computer Vision and Graphics International Conference ICCVG 2008), Springer Verlag, 451–460, ISBN 978-3-642-02344-6.

Mazurek, P. (2007). Estimation Track–Before–Detect motion capture systems state space spatial component. *Lecture Notes in Computer Science* Vol. 4673 (Computer Analysis of Images and Patterns), Springer Verlag, 149-156, ISBN 978-3-540-74271-5.

Michalewicz, A. (2009). *Genetic Algorithms + Data Structures = Evolution Programs*, Springer, ISBN 3540606769.

Otake, M. (2010). *Electroactive Polymer Gel Robots, Modelling and Control of Artificial Muscles*, Springer, ISBN 3540239553.

Sanderson, A.C. & Weiss, L.E. (2008). Adaptive visual servo control of robots, In: *Robot Vision*, Pugh A., (Ed.), 107–116, IFS, ISBN 0903608324.

Spears, W.M. (2000). *Evolutionary Algorithms. The Role of Mutation and Recombination*, Springer, ISBN 3540669507.

Verrelst, B. (2005). *A dynamic walking biped actuated by pleated pneumatic artificial muscles: Basic concepts and control issues* PhD Dissertation, Vrije Universiteit Brussel, http://lucy.vub.ac.be/publications/PhD_Verrelst.pdf.

Wallace, G.G., Spinks, G.M., Kane–Maguire, L.A.P., Teasdale, P.R. (2009). *Conductive Electroactive Polymers. Intelligent Polymer Systems*, CRC Press, ISBN 1420067095.

4

Fuzzy Modelling Stochastic Processes Describing Brownian Motions

Anna Walaszek-Babiszewska
Opole University of Technology
Poland

1. Introduction

Wiener process, as a special mathematical model of Brownian motions, has been investigated and modelling in many probabilistic examples. In the topic literature it is easy to find many procedures of numeric probabilistic simulations of the Wiener process. Fuzzy modelling does not give us more accurate models than probabilistic modelling. Fuzzy knowledge-based modelling allows to determine linguistic description of non-precise relationships between variables and to derive the reasoning procedure from non-crisp facts. More over, using the notions of probabilities of fuzzy events, it is possible to determine a frequency of a conclusion as well as its expected value.

Wiener process and a random walk are very often used for modelling phenomena in physics, engineering and economy. In the area of robot control theory these processes can represent some time-varying parameters of the environments where the object of control operates. Fuzzy models of these processes can constitute a part of a fuzzy model of a tested complex system.

In paragraph 2. of this chapter, the mathematical descriptions of Brownian motions has been reminded, according to the theory of probability and stochastic processes. Some basics of fuzzy modelling has been presented in paragraph 3., to show the method of creating the knowledge base and rules of reasoning. Attention is focused on identification techniques for building empirical probabilities of fuzzy events from input-output data. Exemplary calculations of knowledge bases for real stochastic processes, as well as, some remarks on future works have been presented in paragraphs 4 and 5.

2. Mathematical models of Brownian motion

The Brownian motion it is well known in physics, a random movement of a particle suspended in a liquid or a gas. The name of the movement is given after the botanist Robert Brown (1827), who was studying the movement of pollen grains suspended in water. There are many similar phenomena, where the time evolutions of the object depend on stochastic, microscopic contacts (collisions) with elements of the surrounded system. In mathematics, many models describing Brownian motion are well known and applied, e.g. the random walk stochastic process, Wiener stochastic process, Langevin stochastic differential equation, general diffusion equations and others.

Observations of the microscopic particle behavior show, that at any time step, the particle is changing its position in the space, according to collisions with liquid particles. Crashes of particles are frequent and irregular. It is usually assumed by mathematicians, that the displacements of the particle $Z_1, Z_2, ..., Z_n, ...$ at particular time steps, are independent, identically distributed random variables. The stochastic process $\{Z_i, i = 1, 2, ..., n., ...\}$ is named *random walk*.

In macroscopic scale, if the time between two observations of the particle, $t - \tau$, is larger than the time between successive crashes, then the increment of the particle positions, $X_t - X_\tau$, is a sum of many small displacements, $X_t - X_\tau = \sum\limits_{i=1,...,k} Z_i$. Since the increments

$X_t - X_\tau$ constitute sums of independent, identically distributed random variables, they are normal distributed random variables.

In mathematics, scalar stochastic process $\{X_t, 0 \leq t < \infty\}$, is the *Wiener process* if and only if

i.   increments $X_t - X_\tau$, $0 \leq \tau < t < \infty$ are homogeneous (stationary) and independent for disjoint time intervals,

ii.   the initial condition, $P(X_0 = 0) = 1$, is fulfilled,

iii. trajectories of the process $\{X_t, 0 \leq t < \infty\}$ are continuous (almost surely),

iv.   random variables X_t, are normal distributed, with the probability density function

$$f(t, x) = \frac{1}{\sqrt{2\pi t \sigma^2}} \exp\left(-\frac{x^2}{2t\sigma^2}\right). \tag{1}$$

Wiener process is also known as the *Brownian motion process* (Fisz, 1967; van Kampen, 1990; Kushner, 1983; Sobczyk, 1991).

The increments, $X_t - X_\tau$, $0 \leq \tau < t < \infty$, are normal distributed random variables with the expected value and variance as follows:

$$E(X_t - X_\tau) = 0, \ D^2(X_t - X_\tau) = (t - \tau)\sigma^2. \tag{2}$$

Random variables $X_{t_1}, ..., X_{t_n}$, where

$$X_{t_n} = X_{t_1} + (X_{t_2} - X_{t_1}) + ... + (X_{t_n} - X_{t_{n-1}}), \tag{3}$$

are also normal distributed with parameters:

$$E(X_{t_k}) = 0, \ D^2(X_{t_k}) = t_k \sigma^2, \ k=1,2,...,n; \tag{4}$$

and with a non-zero covariance matrix.

If $\sigma^2 = 1$ then $\{X_t, 0 \leq t < \infty\}$ is the *standard Wiener process*.

Probability, that a particle occurs in some interval $[a, b]$, at the moment t, is given by the relationship

$$\Pr\{X_t \in [a,b]\} = \int\limits_a^b f(t,x)dx = \frac{1}{\sqrt{2\pi t \sigma^2}} \int\limits_a^b \exp\left(-\frac{x^2}{2t\sigma^2}\right)dx. \tag{5}$$

For any $t_1 \leq t_2$ probability density function of the random vector variable (X_{t_1}, X_{t_2}) can be obtained as follows:

$$f(t_1, x_1; t_2, x_2) = f(t_2, x_2 / t_1, X_{t_1} = x_1) f(t_1, x_1) \tag{6}$$

where

$$f(t_2, x_2 / t_1, X_{t_1} = x_1) = \frac{1}{\sqrt{2\pi\sigma(t_2 - t_1)}} \exp\left(-\frac{(x_1 - x_2)^2}{2(t_2 - t_1)\sigma^2}\right) \tag{7}$$

is a conditional probability density function, and $f(t_1, x_1)$ is given by formula (1) with parameters: $E(X_{t_1}) = 0$, $D^2(X_{t_1}) = t_1\sigma^2$.

For any $t_1 \leq t_2 \leq \ldots \leq t_n$ probability density function of the multidimensional random vector $(X_{t_1}, \ldots, X_{t_n})$ can be obtained, taking into account Markov features of the process and using (3), as follows (Sobczyk, 1991):

$$f(t_n, x_n, \ldots, t_1, x_1) = \Pi_{i=1,\ldots,n} \frac{1}{\sqrt{2\pi\sigma(t_i - t_{i-1})}} \exp\left(-\sum_{i=1,\ldots,n} \frac{(x_i - x_{i-1})^2}{2(t_i - t_{i-1})\sigma^2}\right). \tag{8}$$

Stochastic vector process $\{[X_1(t), \ldots, X_n(t)],\ 0 \leq t < \infty\}$ is called the nD *stochastic Wiener process* if its every component, $\{X_i(t),\ 0 \leq t < \infty\}$, $i=1,\ldots,n$, is the scalar stochastic Wiener process and particular scalar stochastic processes are independent.

As an example of the 3D stochastic Wiener process we can show three coordinates of the Brownian particle trajectory.

The Wiener process is also the special diffusion stochastic process, fulfilling the Fokker-Planck diffusion equation

$$\frac{\partial f(x,t)}{\partial t} = \gamma \frac{\partial^2 f(x,t)}{\partial x^2}, \quad \lim_{\Delta t \to 0, \Delta x \to 0} \frac{(\Delta x)^2}{\Delta t} = const = 2\gamma \tag{9}$$

where the solution is given by the normal probability density function, and the diffusion coefficient is equal to $\gamma = \sigma^2 / 2$ (van Kampen, 1990).

In macroscopic scale, in physics and in industrial practice, the probability value $f(x)dx$ that scalar variable X assumes its value from the interval $[x, x+dx]$ is equivalent to the quotient n/N (concentration of particles), where n defines the power of subset of particles, whose feature X determines the value over the interval $[x, x+dx]$, and N is the population size. This idea is consistent with Einstein's experiments who considered collective motion of Brownian particles. He assumed that the density (concentration) of Brownian particles $\rho(x,t)$ at point x at time t, met the following diffusion equation:

$$\frac{\partial \rho(x,t)}{\partial t} = \gamma \frac{\partial^2 \rho(x,t)}{\partial x^2} \tag{10}$$

where γ is a diffusion coefficient. The solution has the known exponential form. From the analytical form of the solution the second central moment of the displacement is expressed as

$$E(X_t^2) = 2\gamma t \tag{11}$$

Diffusion coefficient, γ, has been expressed by Einstein as a function of macro- and microscopic parameters of the fluid and particles, respectively. Einstein confirmed statistical character of the diffusion law (cited by van Kampen, 1990).

3. Fuzzy knowledge representation of the 'short memory' stochastic process

3.1 Stochastic process with fuzzy states

Let $X(t)$ be a 'short memory' stochastic process, the family of time-dependent random variables, where $X \in \chi \subset R$, $t \in T \subset R$ and B is the Borel σ-field of events. Let p be a probability, the normalized measure over the space (χ, B).

Moreover, assume that according to human experts' suggestions, in the universe of process values, the linguistic random variable has been determined with the set of linguistic values, $L(X)=\{LX_i\}$, $i=1,2,\ldots,I$ e.g. $L(X)=\{low, middle, high\}$, according to Zadeh's definition of the linguistic variable (Zadeh, 1975). The meanings of the linguistic values are represented by fuzzy sets A_i, $i=1,2,\ldots,I$ determined on χ by their membership functions, $\mu_{A_i}(x): \chi \to [0,1]$, which are Borel measurable functions, fulfilling the condition

$$\sum_{i=1}^{I} \mu_{A_i}(x) = 1, \ \forall x \in \chi. \tag{12}$$

According to above assumptions, the probability distribution of linguistic values of the process $X(t)$ can be determined as follows

$$P(X_t) = \{P(A_i)\}, i = 1,2,\ldots,I, \tag{13}$$

based on Zadeh's definitions of the probability of fuzzy events (Zadeh, 1968)

$$P(A) = \int_{\chi \subseteq R^n} \mu_A(x)dp. \tag{14}$$

The following conditions must be fulfilled

$$0 \le P(A_i) \le 1, i=1,2,\ldots,I; \quad \sum_{i=1}^{I} P(A_i) = 1. \tag{15}$$

Let now $t = t_1$, $t = t_2$, $t_2 > t_1$ be fixed, so the stochastic process at that moments is represented by two random variables $(X(t_1), X(t_2))$. Assume, that (χ^2, B, p) is a probability space, where $\chi^2 \subseteq R^2$, B is the Borel σ-field of events and p is a probability, the normalized measure over (χ^2, B). The assumptions mean that the probability distribution $p(x_{t_1}, x_{t_2})$ over the realizations (X_{t_1}, X_{t_2}) exists.

Let also two linguistic random variables (linguistic random vector) (X_{t_1}, X_{t_2}) be generated in χ^2, taking simultaneous linguistic values $LX_i \times LX_j$, $i,j=1,2,\ldots,I$; corresponding collection of fuzzy events $\{A_i \times A_j\}_{i,j=1,\ldots,I}$ is determined on χ^2 by membership functions

$\mu_{A_i \times A_j}(x_{t_1}, x_{t_2})$, $i,j=1,2,...,I$. Membership functions for joint fuzzy events $A_i \times A_j$ should fulfill

$$\sum_i \sum_j \mu_{A_i \times A_j}(x_{t_1}, x_{t_2}) = 1, \ \forall (x_{t_1}, x_{t_2}) \in \chi^2 . \tag{16}$$

The joint 2D probability distribution of linguistic values (fuzzy states) of the stochastic process $X(t)$ is determined by the joint probability distribution of the linguistic random vector (X_{t_1}, X_{t_2})

$$P(X_{t_1}, X_{t_2}) = \{P(A_i \times A_j)\}_{i,j=1,2,...,I} \tag{17}$$

calculated according to

$$P(A_i \times A_j) = \int_{(x_{t_1}, x_{t_2}) \in \chi^2} \mu_{A_i \times A_j}(x_{t_1}, x_{t_2}) dp \tag{18}$$

and fulfilling

$$0 \leq P(A_i \times A_j) \leq 1, \ \forall i,j = 1,...,I \ \text{ and } \ \sum_{i=1}^{I} \sum_{j=1}^{I} P(A_i \times A_j) = 1 \tag{19}$$

(Walaszek-Babiszewska, 2008, 2011). From the joint probability distribution (17), the conditional probability distribution of the fuzzy transition

$$P[(X_{t_2} = A_j) / (X_{t_1} = A_i)] , j=1,2,...,I; \ i=const \tag{20}$$

can be determined according to

$$\begin{aligned} \{ \ P[(X_{t_2} / (X_{t_1} = A_i)] \ \}_{j=1,...,I} = \\ = \frac{\{P(X_{t_2} = A_j, X_{t_1} = A_i) \ \}_{j=1,...,I}}{P(X_{t_1} = A_i)} . \end{aligned} \tag{21}$$

The following relationships should be fulfilled for the conditional distributions of fuzzy states (probability of the transitions)

$$\sum_{j=1,...,I} P[(X_{t_2} = A_j / (X_{t_1} = A_i)] = 1; \ i = const. \tag{22}$$

3.2 Rule based fuzzy model

The proposed model of the stochastic process, formulated into fuzzy categories, for two moments t_1, t_2, $t_2 > t_1$, is a collection of file rules, in the following form (Walaszek-Babiszewska, 2008, 2011):

$$\forall A_i \in L(X) , i=1,...,I$$

$$R^{(i)}: w_i[If\,(X_{t_1}\ is\,A_i)]\,Then\,(X_{t_2}\ is\,A_1\,)w_{1/i}$$

$$\text{-----------}$$

$$Also(X_{t_2}\ is\ A_j)w_{j/i} \tag{23}$$

$$\text{-----------}$$

$$Also(X_{t_2}\ is\ A_J\,)w_{J/i}$$

or as a collection of the elementary rules in the form

$$\forall A_i \in L(X)\,,\ \forall A_j \in L(X)\,,\ i,j=1,2,\dots,I$$

$$R^{(i,j)}: w_{ij}[If\,(X_{t_1}\ is\ A_i)\ Then\ (X_{t_2}\ is\ A_j\,)] \tag{24}$$

where the weights w_i, $w_{j/i}$, w_{ij} represent probabilities of fuzzy states, determined by (13) – (15), (20) – (22) and (17) - (19), respectively. The weights stand for the frequency of the occurrence of fuzzy events in particular parts of rules and show the probabilistic structure of the linguistic values of the linguistic random vector $\left(X_{t_1},X_{t_2}\right)$. The weights do not change logic values of the conditional sentences.

3.3 Reasoning procedures
Considering reasoning procedure, we assume that some non-crisp (vague) observed value of the stochastic process at moment t_1 is known and equal to $X_{t_1} = A'$, , or some crisp value $X_{t_1} = x^*_{t_1}$ of the stochastic process at moment t_1 is given. Then, the level of activation of the elementary rule (24) is determined according to one of the following formulas

$$\tau_i = \max_x \min[\mu_{A_i}(x),\mu_{A'}(x)]\,, \tag{25}$$

$$\tau_i = \mu_{A_i}(x^*_{t_1})\,,\ i=1,\dots,I\,, \tag{26}$$

repectively (Yager & Filev, 1994; Hellendoorn & Driankov, 1997). The conclusion according to the generalized Mamdani-Assilian's type interpretation of fuzzy models has the following form

$$\mu_{A'_{j/i}}(x_{t_2}) = T(\tau_i,\mu_{A_j}(x_{t_2}))\,,\,j=1,\dots,I;\,i=\text{const}; \tag{27}$$

thus the conclusion derived based on logic type interpretation of fuzzy models is as follows

$$\mu_{A'_{j/i}}(x_{t_2}) = I(\tau_i,\mu_{A_j}(x_{t_2}))\,,\,j=1,\dots,I;\,i=\text{const}, \tag{28}$$

where T denotes a t-norm and I means the implication operator. Aggregation of the conclusions from particular rules is usually computed by using any s-norm operator (Yager & Filev, 1994; Hellendoorn & Driankov, 1997).
Weights of rules, representing the probability of a fuzzy event in antecedent (w_i), as well as, the conditional probability of a fuzzy event at the consequence part ($w_{j/i}$), can be used to determine probabilistic characteristics of the conclusion. It is worth to note, that fuzzy

conclusions (27) and (28) represent some functions, $\varphi[L(X)]$, of linguistic values of the linguistic random variable X_{t_2}. The fuzzy expected value of the following prediction,

$$E\{(X_{t_2}\ is\ \varphi(A_j))\ /\ [X_{t_1}\ is\ A']\} = \bar{A}, \qquad (29)$$

computed as the aggregated outputs of all active i-th file rules, can be determined by the following formula (Walaszek-Babiszewska, 2011)

$$\mu_{\bar{A}}(x_{t_2}) = \sum_i w_i \mu_{A_i'}(x_{t_2}) = \sum_i w_i \sum_j w_{j/i}\mu_{A'_{j/i}}(x_{t_2}). \qquad (30)$$

where membership functions, $\mu_{A'_{j/i}}$, of the conclusions from elementary rules are given by (27) or (28), depending on the type of input data and the interpretation of a fuzzy model. Also, it is possible to determine probability of the fuzzy conclusion, taking into account a marginal probability distribution $P(X_{t_2})$ of the output linguistic random variable.

4. Creating fuzzy models of stochastic processes - Exemplary calculations

4.1 Fuzzy model of the stochastic time-discrete increments
First example show the fuzzy representation of the simplest form of the considered above stochastic processes, the one-dimensional time-discrete stochastic process of the increments, $\Delta X_t = X_t - X_{t-1}$. The increments, at given t, are normal distributed random variables, so it is useful to use the standard normal probability distribution function, over the domain of the process values, $\Delta X \in [-3, 3] = \chi \subset R$ (Table 1). Linguistic random variable, Y_t , with the

$x \in [a,b]$	$p(x)$	Fuzzy sets (events)					Probability of fuzzy events $P(Y)$	
		$\mu_{NH}(x)$	$\mu_{NL}(x)$	$\mu_Z(x)$	$\mu_{PL}(x)$	$\mu_{PH}(x)$		
[-3, -2.5)	0.00486	1	0				$P(NH)=$ 0.014	
[-2.5, -2)	0.01654	0.5	0.5					
[-2, -1.5)	0.044057	0	1				$P(NL)=$ 0.220	
[-1.5, -1)	0.091848		1	0				
[-1, -0.5)	0.149882		0.5	0.5				
[-0.5, 0)	0.191463		0	1				$P(Z)=$ 0.532
[0, 0.5)	0.191463			1	0			
[0.5, 1)	0.149882			0.5	0.5			
[1, 1.5)	0.091848			0	1		$P(PL)=$ 0.220	
[1.5, 2)	0.044057				1	0		
[2, 2.5)	0.01654				0.5	0.5	$P(PH)=$ 0.014	
[2.5, 3]	0.00486				0	1		

Table 1. Probability function of random variable X_t , fuzzy sets representing linguistic values $L\{Y\}$ of the linguistic random variable Y_t and probability distribution $P(Y)$

name *'Increment at moment t'* has been assumed, with the set of its linguistic values: $L(Y)$={*negative high NH, negative low NL, zero Z, positive low PL, positive high PH*}. The linguistic values are represented by respective fuzzy sets. The probability distribution of the linguistic random variable, $P(Y)$, calculated according to (13) - (15), has been presented in Table 1.

Also, the second linguistic random variable, Y_{t-1}, with the name *'Increment at moment t-1'* has been determined with the same set of linguistic values $L(Y)$. Increments of the tested process are independent random variables, so conditional probabilities (probabilities of transitions) fulfill the relationship: $P(Y_t / Y_{t-1}) = P(Y_t)$.

The fuzzy knowledge base for the short memory stochastic process consists of the following, five file rules (25 elementary rules) with respective probabilities (according to Table 1.):

$$R^1: 0.014(\text{If } Y_{t-1} \text{ is NH) Then } (Y_t \text{ is NH})0.014$$

$$\text{Also } (Y_t \text{ is NL) } 0.220$$

$$\text{Also } (Y_t \text{ is Z)) } 0.532$$

$$\text{Also } (Y_t \text{ is PL) } 0.220$$

$$\text{Also } (Y_t \text{ is PH) } 0.014;$$

$$R^2: 0.220 \text{ (If } Y_{t-1} \text{ is NL) Then } (Y_t \text{ is NH) } 0.014$$

-- (31)

$$R^5: 0.014(\text{If } Y_{t-1} \text{ is PH) Then } (Y_t \text{ is NH) } 0.014$$

$$\text{Also } (Y_t \text{ is PH) } 0.014.$$

In the created rule base of the stochastic process, the same probability distributions for random variables, ΔX_t and ΔX_{t-1}, have been assumed. It is result of the simplification, under the assumption of a constant time interval $\Delta t = 1$.

4.2 Exemplary fuzzy models constructed based on realizations of stochastic processes

4.2.1 Fuzzy model constructed based on data of a floating particle

In the object literature the problem of the fulfilling the Wiener process assumptions by empirical data is often raised, e.g. the expected values of empirical increments are non-zero or increments do not fulfill the criterion of probabilistic independence. These facts have been also observed based on data representing increments of one coordinate, $\Delta x_t = x_t - x_{t-1}$, $\Delta x \in [-3.3, 3.3] = \chi \subset R$, describing the behavior of the particle floating in some liquid. It was assumed, that data stand for the realization of a certain stochastic process $Y(t)$. Also, the linguistic random variable Y_t has been determined, with the name *'Increment at moment t'*.

The set of the linguistic values, $L(Y)$={*negative high NH, negative low NL, positive low PL, positive high PH*} has been assumed. In domain χ of the process values, the linguistic values are represented by respective fuzzy sets. Also, second linguistic random variable, Y_{t-1}, with the name '*Increment at moment t-1*' has been determined with the same set of linguistic values $L(Y)$. For the tested process, the criterion of independent increments is not fulfilled, thus, the conditional probabilities (probabilities of transitions) $P(Y_t / Y_{t-1})$ should be found.

The empirical joint probability of two linguistic random variables, $P(Y_t, Y_{t-1})$, has been calculated according to (16) - (19), based on the joint probability of numeric values of pairs, $p(\Delta x_{t-1}, \Delta x_t)$, as well as, the assumed fuzzy events, representing the linguistic values $L\{Y_t\} = \{NH, NL, PH, PL\}$ (Table 2). Marginal probability $P(Y_{t-1})$ is presented at the last row of the table. It is not a symmetrical distribution, the highest value of the probability, 0.39251, it is a probability that increments take the linguistic value '*Positive Low*'.

Conditional probabilities $P(Y_t / Y_{t-1})$, calculated according to (20) - (22) and presented in Table 3, may be treated as the transitions probabilities from fuzzy states {NH, NL, PL, PH} at moment t-1 to the particular fuzzy states at moment t.

$L\{Y_t\}$	$P(Y_t, Y_{t-1})$			
	$L\{Y_{t-1}\}$			
	NH	NL	PL	PH
PH	0.0178	0.05355	0.10702	0.03572
PL	0.0060	0.06555	0.19024	0.09532
NL	0.0953	0.05955	0.02975	0.06555
NH	0.04765	0.03570	0.0655	0.0298
$P(Y_{t-1})$	0.16675	0.21435	0.39251	0.22639

Table 2. Joint probability distribution, $P(Y_t, Y_{t-1})$, of linguistic random variables representing empirical set of increments at moments t and t-1

$L\{Y_t\}$	$P(Y_t / Y_{t-1})$			
	$L\{Y_{t-1}\}$			
	NH	NL	PL	PH
PH	0.10675	0.24983	0.27266	0.15778
PL	0.03600	0.30581	0.48466	0.42104
NL	0.57150	0.27781	0.07580	0.28955
NH	0.28575	0.16655	0.16688	0.13163
$\sum P(Y_t / Y_{t-1})$	1.00000	1.00000	1.00000	1.00000

Table 3. Conditional probability distributions $P(Y_t / Y_{t-1})$ of linguistic random variables

The fuzzy knowledge base of the behavior of some particle, determined by changes of its coordinate Y_t , consists of four file rules (20 elementary rules) with respective probabilities, as follows:

$$R^1: 0.16675 \text{ (If } Y_{t-1} \text{ is NH) Then } (Y_t \text{ is NH) } 0.28575$$

$$\text{Also } (Y_t \text{ is NL) } 0.57150$$

$$\text{Also } (Y_t \text{ is PL) } 0.036$$

$$\text{Also } (Y_t \text{ is PH) } 0.10675;$$

$$R^2: 0.21435 \text{ (If } Y_{t-1} \text{ is NL) Then } (Y_t \text{ is NH) } 0.16655$$

$$\text{---} \quad (32)$$

$$R^4: 0.22639 \text{(If } Y_{t-1} \text{ is PH) Then } (Y_t \text{ is NH) } 0.13163$$

$$\text{-----------------------------}$$

$$\text{Also } (Y_t \text{ is PH) } 0.15778.$$

Probabilities (weights) at the consequent stand for transitions probabilities.

4.2.2 Fuzzy model of the stochastic increments observed in some technological situation

In a certain technological situation some parameter of a non-homogeneous grain material was measured at discrete moments (Figure 1). It is assumed that observed values $X(n)$, $n=1,...,400$ represent realization of a certain stochastic process whose variance is high and changes are very quick. For human experts, engineers of the technological process, it is very

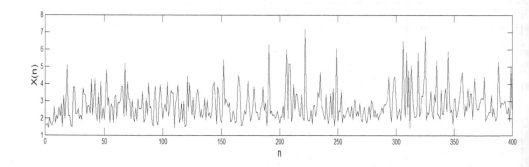

Fig. 1. Realization of the stochastic process $X(n)$

important to recognize a probabilistic character of the changes, especially the changes determined in linguistic categories, like: *Positive Big, Positive Small, Zero, Negative Small, Negative Big*. To determine characteristic of the process with fuzzy states, first, we have calculated the increments

$$DX(n)=X(n)-X(n-1) \tag{33}$$

and a joint probability distribution $p(DX(n),DX(n-1))$ of non-fuzzy values of the process. The range of the increments values, a real number interval [-4.8, 4.8], has been divided into 14 disjoint intervals and the frequency of the occurrence of measurements in particular intervals has been determined. The disjoint intervals have been used for the description of membership functions of particular linguistic values of the set $L\{DX(n)\}=\{NB,NS,Z,PS,PB\}$.

The empirical joint probability distribution of the linguistic random variables, $P(L\{DX(n-1)\}, L\{DX(n)\})$, has been calculated and presented in Table 4. In the last row, the marginal probability values of one linguistic random variable, are presented. It is almost symmetrical distribution, with the highest value of the probability, 0.6114, for the linguistic value of increments equal to *'Zero'*. Conditional probability distributions for particular linguistic values of the variable $DX(n)$ have been also calculated and they represent weights of particular consequent parts of the rule-base fuzzy model (34). The model of the knowledge base consists of the following five file rules with weights:

R1: 0.6114 IF $(DX(n-1)$ IS $Z)$ THEN $(DX(n)$ IS $Z)$ 0.6450

ALSO $(DX(n)$ IS $NS)$ 0.1945

ALSO $(DX(n)$ IS $PS)$ 0.1462

ALSO $(DX(n)$ IS $PB)$ 0.0113

ALSO $(DX(n)$ IS $NB)$ 0.0030

R2: 0.2123 IF $(DX(n-1)$ IS $NS)$ THEN $(DX(n)$ IS $Z)$ 0.6739

ALSO $(DX(n)$ IS $PS)$ 0.2032

ALSO $(DX(n)$ IS $NS)$ 0.1114

ALSO $(DX(n)$ IS $PB)$ 0.0111

ALSO $(DX(n)$ IS $NB)$ 0.0004

R3: 0.1474 IF $(DX(n-1)$ IS $PS)$ THEN $(DX(n)$ IS $NS)$ 0.4258   (34)

ALSO $(DX(n)$ IS $Z)$ 0.4181

ALSO $(DX(n)$ IS $NB)$ 0.0791

ALSO $(DX(n)$ IS $PS)$ 0.0714

ALSO ($DX(n)$ IS PB) 0.0056

R4: 0.0184 IF ($DX(n-1)$ IS NB) THEN ($DX(n)$ IS Z) 0.6044

ALSO ($DX(n)$ IS PS) 0.2363

ALSO ($DX(n)$ IS NS) 0.1263

ALSO ($DX(n)$ IS PB) 0.0330

R5: 0.0105 IF ($DX(n-1)$ IS PB) THEN ($DX(n)$ IS NB) 0.4762

ALSO ($DX(n)$ IS NS) 0.4286

ALSO ($DX(n)$ IS Z) 0.0952.

	$P(L\{DX(n-1)\}, L\{DX(n)\})$				
	$L\{DX(n-1)\}$				
$L\{DX(n)\}$	NB	NS	Z	PS	PB
PB	0.0051	0.0046	0.0010	0.0	0.0
PS	0.0115	0.0620	0.0609	0.0104	0.0008
Z	0.0017	0.1192	0.3953	0.0895	0.0068
NS	0.0001	0.0235	0.1430	0.0431	0.0023
NB	0	0.0030	0.0112	0.0044	0.0006
$P(L\{DX(n-1)\})$	0.0184	0.2123	0.6114	0.1474	0.0105

Table 4. Joint empirical probability distribution of two linguistic random variables representing increments

To determine the predicted value $DX(n)=b^*$, for given value (crisp or fuzzy) $DX(n-1)=a^*$, the reasoning procedure, described in 3.3 is used, e.g. for $DX(n-1)=1.55$, predicted value is approximated as equal to $DX(n)=0.30538$. This value depends on many parameters of the fuzzy model and the reasoning procedure. It is very useful to create the computing system with many options of changing the reasoning parameters. In Fig. 2 the predicted, mean values of the increments has been underlined by thick line.

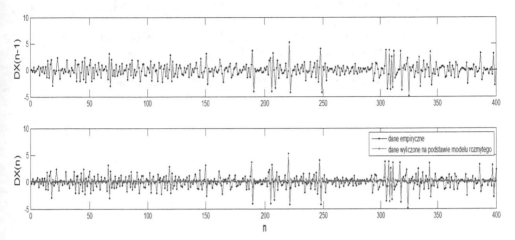

Fig. 2. Realization of the stochastic processes of increments $DX(n)$, $DX(n-1)$ and the predicted mean value

5. Conclusion and future works

In this chapter the new approach to fuzzy modelling has been presented. Knowledge base in the form of weighted fuzzy rules represents in the same time the probability distribution of the fuzzy events occurring in the statements. Considered examples show the creating a few simple models of stochastic increments processes. In the future, in modelling the Wiener process, the time dependent probability of the increments should be taken into account.

6. References

Fisz, M. (1967). *Probability and Statistics Theory* (in Polish), PWN, Warsaw, Poland

Hellendoorn, H. & Driankov, D. (Eds.), (1977). *Fuzzy Model Identification; Selected Approaches*, Springer, ISBN 3-540-62721-9, Berlin, Germany

Kushner, H. (1983). *Introduction to Stochastic Control*, PWN (Polish edition), ISBN 83-01-02212-4, Warsaw, Poland

Sobczyk K. (1996). *Stochastic Differential Equations*, WNT (Polish edition), ISBN 83-204-1971-9, Warsaw, Poland

Van Kampen, N.G. (1990). *Stochastic Processes in Physics and Chemistry*, PWN (Polish edition), ISBN 83-01-09713-2, Warsaw, Poland

Walaszek-Babiszewska, A. (2008). Probability Measures of Fuzzy Events and Linguistic Fuzzy Modelling – Forms Expressing Randomness and Imprecision, In: *Advances on Artificial Intelligence, Knowledge Engineering and Data Bases*, L.A. Zadeh, J. Kacprzyk et al. (Eds.), *Proceedings of the 7th WSEAS International Conference AIKED'08* pp.207-213, ISBN 978-960-6766-41-1, Cambridge, UK, February 20-22, 2008

Walaszek-Babiszewska, A. (2011). *Fuzzy Modelling in Stochastic Environment; Theory, knowledge bases, examples*, LAP Lambert Academic Publishing, ISBN 978-3-8454-1022-7, Saarbrucken, Germany

Yager, R. & Filev, D.P. (1994). *Essentials of Fuzzy Modelling and Control.* John Wiley and Sons, New York, USA

Zadeh, L.A. (1968). Probability Measures of Fuzzy Events. *Journal of Mathematical Analysis and Applications*, Vol.23, No.2, (August 1968), pp. 421-427

Zadeh, L.A. (1975). The Concept of a Linguistic Variable and its Applications to Approximate Reasoning –I. *Information Sciences*, Vol.8, pp. 199-249

5

Heuristic Optimization Algorithms in Robotics

Pakize Erdogmus[1] and Metin Toz[2]
[1]Duzce University, Engineering Faculty,
Computer Engineering Department
[2]Duzce University, Technical Education Faculty,
Computer Education Department, Duzce
Turkey

1. Introduction

Today, Robotic is an essential technology from the entertainment to the industry. Thousands of articles have been published on Robotic. There are various types of robots such as parallel robots, industrial robots, mobile robots, autonomous mobile robots, health-care robots, military robots, entertainment robots, nano robots and swarm robots. So, this variety brings a lot of problems in Robotic. Inverse kinematic for serial robots, forward kinematic for parallel robots, path planning for mobile robots and trajectory planning for industrial robots are some of the problems in Robotic studied a lot. Some of the problems are solved easily with some mathematical equations such as forward kinematic problem for serial robots and inverse kinematic problem for parallel robots. But the problems consisting of nonlinear equations and higher order terms can't be solved exactly with the classical methods. The forward kinematics problem of the 6 degrees of freedom (DOF) 6x6 type of Stewart Platform can be given as an example to such problems. It has been shown that there are 40 distinct solutions for this problem (Raghavan, 1993). Some of the unsolved problems with classical methods are optimization problems and heuristic optimization techniques are an alternative way for the solution of such problems. The heuristic optimization techniques produce good solutions for the higher order nonlinear problems in an acceptable solution time, when the problems aren't solved with the classical methods (Lee & El-Sharkawi, 2008).

In this chapter, first in Section 2, optimization is shortly described. The introduction and some well-known heuristic algorithms including, Genetic Algorithms(GA), Simulated Annealing(SA), Particle Swarm Optimization(PSO) and Gravitational Search Algorithm(GSA) are reviewed in Section 3. In Section 4, two well-known optimization problems in Robotic are solved with GA and PSO.

2. Optimization

Optimization is of great importance for the engineers, scientists and managers and it is an important part of the design process for all disciplines. The optimal design of a machine, the minimum path for a mobile robot and the optimal placement of a foundation are all optimization problems.

A constrained optimization problem has three main elements; design variables, constraints and objective function/functions. Design variables are independent variables of the objective function and can take continuous or discrete values. The ranges of these variables are given for the problems. Constraints are the functions of design variables and limit the search space. The objective function is the main function dependent on the design variables. If there is more than one objective function, the problem is called multi-objective optimization problem.

Solving an optimization problem means finding the values of the design variables which minimize the objective function within given constraints. So if the objective function is a maximization problem, it is converted to minimization problem multiplying by -1 as seen in the Figure 1. The mathematical model of an optimization problem is given by equation 1.

$$g_i(x_1, x_2, ... x_n, a) \leq 0, \quad i = 1, 2, ..., l$$

Minimize $f_{obj}(x_1, x_2, ... x_n)$, Subject to $h_j(x_1, x_2, ... x_n, b) = 0, \quad j = 1, 2, ..., m$ (1)

$$x_k^l \leq x_k \leq x_k^u, \quad k = 1, 2, ... n$$

Where, x_1, x_2, ...x_n are design variables, n is the number of design variables, l is the number of inequality constraints and m is the number of equality constraints, a and b are constant values. x_k^l and x_k^u are respectively lower and upper bounds of the design variables.

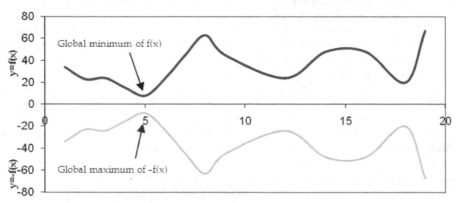

Fig. 1. An objective function conversion to minimization

There are a lot of applications in decision science, engineering, and operations research which can be formulated as constrained continuous optimization problems. These applications include path planning for mobile robots, trajectory planning for robot manipulators, engineering design and computer-aided-design (CAD). Optimal or good solutions to these applications are very important for the system performance, such as low-cost implementation and maintenance, fast execution and robust operation (Wang, 2001).

There are different methods for the solution of the optimization problems. Exact methods and heuristics are two main solution methods. Exact methods find certain solutions to a given problem. But, if the problem size increases, the solution time of the exact methods is unacceptable, because of the fact that solution time increases exponentially when the problem size increases. So, using the heuristics methods for the solution of optimization problems are more practical. (Rashedi et al., 2009). On the other hand, in last decades there has been a great deal of interest on the applications of heuristic search algorithms to solve the such kind of problems.

The main difficulty encountered in the solution of the optimization problem is the local minimums. If there are a lot of local minimums, then both exact methods and heuristics can be trapped in to the local minimums. Since most of the heuristic algorithms developed a strategy to avoid the local minimums, they have been quite popular recently. The local minimums and global minimum of an objective function with one design variable are seen in the Figure 2.

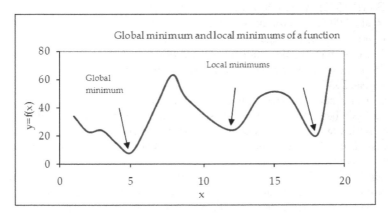

Fig. 2. An objective function with local minimums

3. Heuristic algorithms

Optimization studies drawn attention in 1960's. It was developed a lot of algorithm based on the more sophisticated mathematical background. These algorithms were called exact methods. But, exact methods produced certain solutions only for the limited scope of application. As a result, attention turned to heuristic algorithms. (Fisher et al., 1998). Heuristics algorithms try to find acceptable solutions to the problems using some heuristic knowledge and most of them simulate real life. They use not only pure mathematics, but also algorithms using with basic formulations. While the most important property of heuristic algorithms is that they are designed for the unconstrained optimization problems, they can also be adapted to the constrained optimization problems. The algorithms are designed to find the minimum value of the objective function within the bounds of the constraints . If a solution doesn't satisfy the constraints, this solution is not acceptable, even if the value of the objective function is minimum. So, if there are constraints in a minimization problem, penalty function is added to objective function. The total function is called as fitness function. Namely, if constraints are in feasible region, then there is no penalty and penalty function is equal to zero. If the constraints are not in feasible region, the fitness function is penalized by penalty function.

Heuristic algorithms don't guarantee finding the optimal solutions. They try to find acceptable solutions near to optimum in a reasonable time. These algorithms were studied for both discrete and continuous optimization problems since the 1970's. Researchers have tried to develop adaptive and hybrid heuristics algorithms. By this aim, tens of algorithms were developed in last decade. Heuristic algorithms can be classified as single solution algorithms and population based algorithms. The first category gathers Local Search (LS), Greedy Heuristic (GH) (Feo et al., 1995), Simulated Annealing (SA) (Kirkpatrick et al., 1983), Tabu Search (TS) (Laguna, 1994) and Iterated Local Search (ILS) (Glover & Garry, 2002). The second category, which is more and more studied, groups evolutionary algorithms such as Genetic Algorithms (GA)(Goldberg, 1989), Ant Colony Optimization (ACO) (Dorigo et al, 1996), Artificial Bee Colony (ABC)(Basturk & Karaboga, 2006), Scatter Search (SS) (Marti et al., 2006), Immune Systems (IS) (Farmer et al., 1986), Differential Evolution Algorithms (DEA) (Storn, 1996), Particle Swarm Optimization (PSO) (Keneddy & Eberhart, 1995), Harmony Search (HS) (Geem & Kim, 2001) and Gravitational Search Algorithm (GSA) (Rashedi et al., 2009). Evolutionary algorithms simulate natural phenomena or social behaviors of animals. So, the algorithms can be subcategorized to these criteria. While SA, GSA, HS simulates natural phenomena, ACO, ABC, PSO simulates the social behaviors of animals. The taxonomy of global optimization algorithms can be found in the related literature (Weise, 2008).

During the last years, hybrid optimization approaches have been popular. Firstly, cooperations have been realized between several heuristic algorithms. Later, cooperations between heuristic algorithms and exact methods have been realized. Since they gather the advantages of both types of algorithm, hybrid algorithms give more satisfied results. (Jourdan et al., 2009).

In single solution heuristic algorithms, the algorithm starts with an initial solution called current solution. This solution can be created randomly and consists of the values of design variables of the optimization problems. The objective function value is calculated with these initial values. If the initial solution satisfies the constraints, fitness function will be the same with the objective function. The second solution, called candidate solution, is created with random values close to the initial solution. Between the two fitness functions, the one which has a minimum value is selected as candidate solution. As long as the iteration proceeds, it is expected that algorithm converges to the global minimum value. But algorithms can be trapped to the local minimums. Heuristic algorithms develop different strategies to avoid to get trapped the local minimums. So, exploration and exploitation are of great importance for finding the global minimum value with the heuristics algorithms. In the first iterations, the algorithm searches the global solution space with big steps in order not to get trapped the local minimums. This is called exploration. In the next iterations, the algorithm searches the solution space with small steps in order to find best minimum. This is called exploitation. It is desired to have a balance between exploration and exploitation in the heuristic algorithms.

3.1 Genetic algorithms

GA was developed in Michigan University by Holland and later it got popularity with the efforts of Goldberg. GA is an adaptive global search method that mimics the metaphor of natural biological evolution. It operates on a population of potential solutions by applying the principle of survival of the fittest to achieve an optimal solution.

In the literature, GA is one of the most applied heuristic optimization algorithms. GA has been applied successfully to a huge variety of optimization problems, nearly in all disciplines, such as materials science, aircraft applications, chemistry, construction, seismology, medicine and web applications. GA has also been applied to the problems in Robotic such as the solution of multimodal inverse kinematics problem of industrial robots

(Kalra et al., 2006), trajectory generation for non-holonomic mobile manipulators (Chen & Zalzala, 1997), generation of walking periodic motions for a biped robot (Selene et al, 2011), the solution of the forward kinematics of 3RPR planar parallel manipulator(Chandra & Rolland, 2011), kinematic design optimization of a parallel ankle rehabilitation robot (Kumar et al., 2011).

GA starts to run with a lot of possible solutions according to the initial population which are randomly prepared. Each individual is encoded as fixed-length binary string, character-based or real-valued encodings. Encoding is an analogy with an actual chromosome (Lee & El-Sharkawi, 2008). A binary chromosome represents the design variables with a string containing 0s and 1s. Design variables are converted to binary arrays with a precision p. The bit number is calculated by equation 2 (Lin & Hajela, 1992).

$$B. \; N \geq \log_2(((x_k^u - X_k^l)/p)+1) \tag{2}$$

Where, p is a precision selected by the program designer. x_k^u and x_k^l are respectively upper and lower bounds of the design variables $B.N$ is the number of bits representing design variables. If the number of bits is 4, the binary values of the design variables are represented as 0000, 0001, 0010, ..., 1111.

Initial population is generated randomly after coding the variables. Initial population is represented as a matrix. Typical population size varies between 20 and 300. Each row of the population matrix is called as an individual.

Strings are evaluated through iterations, called generations. During each generation, the strings are evaluated using fitness function. Fitness function measures and evaluates the coded variables in order to select the best fitness strings (Saruhan, 2006). Then, it tries to find optimum solutions by using genetic operators. By this way, the best solutions are selected and worsts are eliminated. GA runs according to unconstrained optimization procedure. So, the constrained continuous optimization problems are transformed into unconstrained continuous optimization problem by penalizing the objective function value with the quadratic penalty function. The total function is called fitness function.

Fitness function (FF) values are calculated for each individual as it is seen in equation 3, 4 and 5. In a minimization problem, penalty function is added to objective function. The fitness function (FF), is the sum of objective function (OF) and the penalty function (PF) consisting constraint functions (CF). After elitism, selection, crossover, and mutation are operated, a new population is generated according to the fitness function values. The future of population depends on the evolutionary rules.

$$OF = f_{obj}(x_1, x_2, ..., x_n) \tag{3}$$

$$PF = \sum_{i=1}^{l} R_i(\max[0, g_i(x_1, x_2, ..., x_n, a)])^2 + \sum_{j=1}^{m} Rj(\max[0, |h_j(x_1, x_2, ..., x_n, b)| - tol])^2 \tag{4}$$

$$FF = OF + PF \tag{5}$$

In the above equations, a and b are constant values, l is the number of inequality constraints, m is the number of equality constraints, n is the number of variables and tol is the tolerance value, for the equality constraints. If h_j's value is smaller than the tolerance value, it is accepted that equality constraints have been satisfied. If the problem variables satisfy the constraints, g_i and (h_j –tol) values will be negative and PF will be zero. R_i and R_j are the penalty coefficients and these values affect the algorithm performance directly.

Selection is to choose the individuals for the generation based on the principle of survival of the best according to the Darwin's theory. The main idea in selection is to create parents from the chromosomes having better fitness function values. Namely the parents are selected from the chromosomes of the first population as directly proportional to their fitness function values. The better fitness function value of a chromosome, the more chance to be a parent. With the selection method, the individuals having worse fitness values disappear in the next generations. The most popular selection methods are roulette wheel and tournament rank. The successful individuals are called parents.

After selection, crossover, which represents mating of two individuals, is applied to the parents. The parents in GA create two children with crossover. There are various kinds of crossover process such as one point crossover, multiple point crossovers and uniform crossover. The crossover operator is applied with a certain probability, usually in the range 0.5-1. This operator allows the algorithm to search the solution space. Namely, exploration is performed in GA with crossover operator. In a basic crossover, the children called offspring are created from mother and father genes with a given crossover probability. Mutation is an another operation in GA and some gens are changed with this operation. But this is not common. Generally mutation rate of binary encoding is specified smaller than 0.1. In contrast to the crossover, mutation is used for the exploitation of the solution space. The last operation is the elitism. It transfers the best individual to the next generation. The evaluation of the previous population with the operators stated above, the new population is generated till the number of generation. Fitness function values are calculated in each new population and the best resulted ones are paid attention among these values. Until the stopping criteria are obtained, this process is repeated iteratively. The stopping criteria may be the running time of the algorithm, the number of generation and for fitness functions giving the same best possible values in a specified time. The pseudo codes of GA are given in Figure 3.

```
init_Popu← Produce initial solution ()
(while stopping criteria not true do)
    for i=1 to generation_number
        calculate fitness function values of init_popu
        choose parents with roulette wheel
        (Parent 1 and Parent 2 are chosen)
        Elitism(Select the best ones in the population)
        Crossover(Create two children for each parent)
        Mutation(Gens are changed with mutation_rate)
        The new ofspring is created
    Next
```

Fig. 3. The pseudo codes of GA

3.2 Simulated annealing

SA uses an analogy of physical annealing process of finding low energy states of solids and uses global optimization method. SA is inspired by Metropolis Algorithm and this approach was firstly submitted to an optimization by Kirkpatrick and his friends in 1983 (Kirkpatrick et al, 1983). Because of the fact that SA is an effective algorithm and it is easy to implement for the difficult engineering optimization problems, it is also popular for last studies in several different engineering areas. SA is applied to various optimization problem in Robotics such as path planning problem solution(Gao & Tian,2007), the generation of the assembly sequences in robotic assembly(Hong & Cho, 1999), trajectory optimization of advanced launch system (Karsli & Tekinalp, 2005), optimal robot arm PID control, the placement of serial robot manipulator and collision avoidance planning in multi-robot.

SA combines local search and Metropolis Algorithm. Algorithm starts with a current solution at initial temperature. At each temperature, algorithm iterates n times, defined by the program designer. In the iterations for each temperature, a candidate solution that is close to the current solution is produced randomly. If candidate solution is better than the current solution, the candidate solution is replaced with the current solution. Otherwise, the candidate solution is not rejected at once. The random number is produced between 0-1 and compared with the acceptance probability P. P is an exponential function as given by equation 6. If produced random number is smaller than P, the worse solution is accepted as current solution for exploration. Search process progresses until stopping criteria is satisfied. Maximum run time, maximum iteration number, last temperature e.g. may be the stopping criteria.

$$P = e^{-\left(\frac{f(x\_cand)-f(x\_curr)}{T}\right)} \tag{6}$$

For an object function with one design variable, **x_cand** is candidate design variable, **x_curr** is current design variable and **f** is object function, **T** is temperature. At high temperatures, P is close to one. Since the most random number is smaller than one, the possibility of bad solution's acceptance is high for exploration in the first steps of the algorithm. T is decreased along the search process. At low temperatures, P is close to zero. Since the random numbers are bigger than zero, the possibility of bad solution's acceptance approach to zero. When P is equal to zero, the process converges to local search method for exploitation. The pseudo code of SA is given in Figure 4.

```
x_curr← Produce initial solution()
T←T0
while stopping criteria not true do
    for i=1 to n
            x_cand←Produce a random solution
            if f(x_cand) < f(x_curr) then
                x_ curr ←x_cand
            else
                'Metropolis Algorithm
                x Produce a random number between (0,1)
                if  x < p ( T, x_curr, x_cand) then
                    x_curr ← x_cand
                end if
            end if
        end
        update(T)
    end while
```

Fig. 4. The pseudo codes of SA Algorithm

Annealing process simulates optimization. Annealing is realized slowly in order to keep the system of the melt in a thermodynamic equilibrium. Convergence to the optimal solution is controlled by the annealing process.

Proper cooling scheme is important for the performance of SA. The proper annealing process is related with the initial temperature, iteration for each temperature, temperature decrement coefficient and stopping criteria. All these criteria can be found in related article (Blum & Roli, 2001). In general, the temperature is updated according to equation 7.

$$T_{k+1} = \alpha T_k \qquad (7)$$

Where, T_{k+1} is the next temperature, T_k is the current temperature and α is the temperature decrement coefficient, generally selected between 0.80 and 0.99.

Even if SA tries to find global minimum, it can be trapped by local minima. So, in order to overcome this drawback and develop the performance of SA, researchers have developed adaptive or hybridized SA with other heuristics, such as fuzzy logic, genetic algorithms, support vector machines, and distributed/parallel algorithms.

3.3 Particle swarm optimization

PSO was introduced by James Kennedy and Russell Eberhart in 1995 as an evolutionary computation technique (Kennedy & Eberhart, 1995) inspired by the metaphor of social interaction observed among insects or animals. It is simpler than GA because PSO has no evolution operators such as crossover and mutation (Lazinca, 2009).

PSO have many advantages such as simplicity, rapid convergence, a few parameters to be adjusted and no operators (Li & Xiao; 2008). So, PSO is used for solving discrete and continuous problems. It was applied to a wide range of applications such as function optimization, Electrical power system applications, neural network training, task assignment and scheduling problems in operations research, fuzzy system control and pattern identification. PSO was also applied to a lot of different Robotic applications. PSO was applied mobile robot navigation (Gueaieb & Miah, 2008), Robot Path Planning (Zargar & Javadi, 2009) and Swarm Robotic Applications and Robot Manipulators applications.

PSO algorithm is initialized with a group of particles. Each particle is a position vector in the search space and its dimension represents the number of design variables. PSO is started with random initial positions and velocities belong to each particle. For each particle, fitness function's value is computed. Global and local best values are updated. Local best value is the best value of the current particle found so far and the global best value is the best value of all local bests found so far. Velocity and positions are updated according to the global best and local bests. Each particle flies through the search space, according to the local and global best values. Along the iterations, particles converge to the global best. The convergence of the particles to the best solution shows the PSO performance. The rate of position change of the ith particle is given by its velocity. In the literature there are different velocity formulas. The first one proposed by Keneddy is given by equation 9. Particles' velocity and positions are updated according to equation 8 and 9.

$$x_i^{k+1} = x_i^k + v_i^{k+1} \qquad (8)$$

$$v_i^{k+1} = v_i^k + \varphi_1 \text{rand}()(p_{best i}^k - x_i^k) + \varphi_2 \text{rand}()(g_{best} - x_i^k)) \qquad (9)$$

Where, k is the iteration number, **rand** is a random number, **pbest**$_i{}^k$ is the best value of ith particle and **gbest** is the global best value of all particles. $x_i{}^k$ and $x_i{}^{k+1}$ are respectively the current position and the next position of the ith particle. $v_i{}^k$ and $v_i{}^{k+1}$ are respectively the current velocity and the next velocity of the ith particle. The next position of a particle is specified by its current position and its velocity.

PSO simulates social swarm's behaviour. The velocity formula consists of three important components. First part presents the past velocity. The particle generally continues in the same direction. This is called habit. The second part $\varphi_1 rand()(pbest_i{}^k - x_i{}^k)$ presents private thinking. This part is also called self-knowledge and the last part $\varphi_2 rand()(gbest - x_i{}^k)$ presents the social part of the particle swarm(Kennedy, 1997). When implementing the particle swarm algorithm, some considerations must be paid attention in order to facilitate the convergence and prevent going to infinity of the swarm. These considerations are limiting the maximum velocity, selecting acceleration constants, the constriction factor, or the inertia constant (Valle et al., 2008).

The velocity and position formula with constriction factor (K) has been introduced by Clerc(Clerc, 1999) and Eberhart and Shi (Eberhart & Shi, 2001) as given by equation 10,11 and 12.

$$v_i^{k+1} = K(v_i^k + \varphi_1 rand()(p_{besti}^{\ k} - x_i^k) + \varphi_2 rand()(g_{best} - x_i^k)) \tag{10}$$

$$x_i^{k+1} = x_i^k + v_i^{k+1} \tag{11}$$

$$K = \frac{2}{\left|2 - \varphi - \sqrt{\varphi^2 - 4\varphi}\right|}, \quad \varphi = \varphi_1 + \varphi_2, \quad \varphi > 4 \tag{12}$$

Where, K is the constriction factor, φ_1 and φ_2 represent the cognitive and social parameters, respectively. **rand** is uniformly distributed random number. φ is set to 4.1. The constriction factor produces a damping effect on the amplitude of an individual particle's oscillations, and as a result, the particle will converge over time. The pseudo code of PSO is given in Figure 5.

```
Generate initial P particle swarm's random positions and velocities
do
   for i=1:P
   Evaluate fitness function for each particle
   If fitness(Pi) <fitness(Gbest) then Gbest=Pi
   If fitness(Pi) <fitness(Pbest(i)) then Pbest(i)=Pi
   Update velocity of the particle i
   Update position of the particle i
   end
Until stopping criteria is not true
```

Fig. 5. The pseudo codes of PSO

3.4 Gravitational search algorithm

GSA was introduced by Esmat Rashedi and his friends in 2009 as an evolutionary algorithm. It is one of the recent optimization techniques inspired by the law of gravity. GSA was originally designed for solving continuous problems. Since it is quite newly developed algorithm, it hasn't been applied to a lot of area. GSA was applied to filter modelling (Rashedi et al., 2011), Post-Outage Bus Voltage Magnitude Calculations (Ceylan et al., 2010), clustering (Yin et al., 2011), Scheduling in Grid Computing Systems (Barzegar et al., 2009).

GSA algorithm is based on the Newtonian gravity law: "Every particle in the universe attracts every other particle with a force that is directly proportional to the product of their masses and inversely proportional to the square of the distance between them" (Rashedi et al., 2009).

In GSA, particles are considered as objects and their performances are measured by their masses. GSA is based on the two important formulas about Newton Gravity Laws given by the equation 13 and 14. The first one is as given by the equation 13. This equation is the gravitational force equation between the two particles, which is directly proportional to their masses and inversely proportional to the square of distance between them. But in GSA instead of the square of the distance, only the distance is used. The second one is the equation of accelaration of a particle when a force is applied to it. It is given by equation 14.

$$F = G\frac{M_1 M_2}{R^2} \tag{13}$$

$$a = \frac{F}{M} \tag{14}$$

G is gravitational constant, M_1 and M_2 are masses and R is distance, F is gravitational force, a is acceleration. Based on these formulas, the heavier object with more gravity force attracts the other objects as it is seen in Figure 6.

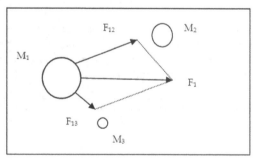

Fig. 6. The Newton Gravitational Force Representation

GSA algorithm initializes the position of the masses(X) given by equation 15.

$$X_i = (x_i^1, x_i^2, \dots, x_i^d, , x_i^n,) \quad i=1,2,3,\dots,N \tag{15}$$

Where, X_i is the position of ith mass. n gives the dimension of the masses. N gives the number of particles. x_i^d represents ith particle position in the dth dimension.

The first velocities of the masses are accepted as zero. Until the stopping criterion (maximum iteration), algorithm evaluates the fitness functions of the objects. After the best and worst values are found for the current iteration, the masses of the particles are updated according to the equation 16, 17 and 18.

$$M_{ai} = M_{pi} = M_{ii} = M_i, \ i = 1, 2, \cdots, N \tag{16}$$

$$m_i(t) = \frac{fitness_i(t) - worst(t)}{best(t) - worst(t)} \tag{17}$$

$$M_i(t) = \frac{m_i(t)}{\sum\limits_{j=1}^{N} m_j(t)} \tag{18}$$

In the equation 17, *fitness_i(t)* represents the fitness value of the *i*th mass at time t, *worst(t)* is the maximum value of all the fitness values and *best(t)* is the minimum value of the all fitness values for a minimization problem. For maximization problems, the opposite of this expression is applied. In GSA, active gravitational mass **Ma**, passive gravitational mass **Mp** and inertial mass **Mi** are accepted as they are equal to each other.

Gravitiational constant which is propotional with the time is also updated according to the iteration number given by the equation 19.

$$G(t) = G_0 e^{-\alpha \frac{t}{T}} \tag{19}$$

Where, α is a user specified constant, T is the total number of iterations, and t is the current iteration. This equation is similar to the SA acceptance probability equation given by the equation 6. According to the formula the gravity force is decreasing by the time. The graviational force and total forces are updated as follows.

$$F_{ij}^{d}(t) = G(t) \frac{M_{pi}(t) M_{aj}(t)}{R_{ij}(t) + \varepsilon} (x_j^d(t) - x_i^d(t)) \tag{20}$$

$$F_i^d(t) = \sum\limits_{j \in Kbest, j \neq i}^{N} rand_j F_{ij}^d(t) \tag{21}$$

$F_{ij}{}^d(t)$ is the gravity force acting on mass *i* from mass *j* at time t, **G(t)** is the gravitational constant at time t, **Mpi** is the passive gravitational mass of object i, **Maj** is the active gravitational mass of object j, ε is a small constant, **R$_{ij}$** is the Euclidean distance between two objects i and j . **F$_i$^{d}(t)** is the force that acts on particle *i* in dimension *d*, **rand$_j$** is a random number between 0 and 1, **Kbest** is the set of first K objects with the best fitness value and biggest mass.The acceleration of an object *i* in *d*th dimension is as follows.

$$a_i^d(t) = \frac{F_i^d(t)}{M_i(t)} \tag{22}$$

The new positions and velocities of the particles are calculated by the equation 23 and 24.

$$v_i^d(t+1) = rand_i \times v_i^d(t) + a_i^d(t) \tag{23}$$

$$x_i^d(t+1) = x_i^d(t) + v_i^d(t) \tag{24}$$

According to the law of motion, particles try to move towards the heavier objects. The heavier masses represent good solutions and they move slowly for the exploitation. The pseudo code of GSA is given in Figure 7.

Initialize the gravitational constant(G) and the positions of N masses(X_i)
$x_i=(x_i^1, x_i^2, \ldots, x_i^d, , x_i^n,)$ i=1,2,3,...,N
for i=1:max_iteration
    Decrease gravitational constant with the equation 19.
    Evaluate the fitness of each object.
    Calculate the masses with the equation 17, 18.
    Compute the total forces with the equation 21.
    Find the accelerations with the equation 22.
    Compute the velocities with the equation 23.
    Compute the positions with the equation 24.
End

Fig. 7. The pseudo codes of GSA

4. PSO and GA applications in robotic

In this section some well-known problems in Robotic were solved with heuristic search algorithms. The first problem is the inverse kinematics problem for a nearly PUMA robot with 6 DOF, and the second problem is the path planning problem for mobile robots.

4.1 The inverse kinematics problem for PUMA robot

Robot kinematics refers to robot motions without consideration of the forces. The forward kinematics is finding the robot's end effector's position and orientation using the given robot parameters and the joint's variables. Also, the inverse kinematics is about finding the robot's joints variables while the robot's parameters and the desired position of the end effector are given (Kucuk & Bingul, 2006). The solution of the robot's forward kinematics is always possible and unique. On the other hand, generally, the inverse kinematics problem has several solutions (Kucuk & Bingul, 2006). The most known and used method for solving robot kinematics problems is Denavit-Hartenberg method. In this method several parameters, namely DH parameters, are defined according to robot's parameters and the replaced coordinate systems as given in Figure 8. These parameters are used to define transformations between the coordinate systems using 4x4 transformation matrices. The multiplication of all the matrices of the robot gives the final robot's end effectors' position and orientation (Kucuk & Bingul, 2006).

The selected robot for this example is a nearly PUMA type robot. Different from the general PUMA robot, the robot used in this example was equipped with an offset wrist. The inverse kinematics of the robot equipped with such a wrist is quite complicated and can be computationally cumbersome (Kucuk & Bingul, 2005). In this example, this problem was solved with GA and PSO. The robot's link parameters and coordinate systems replacements can be seen in Figure 9. Furthermore, according to Figure 9, the DH parameters of the robot can be defined as given in Table 1. Using the DH parameters, the forward kinematics of the robot can be solved easily using robot's transformation matrices. A transformation matrix is a 4x4 matrix and has the form as it is seen in equation 25.

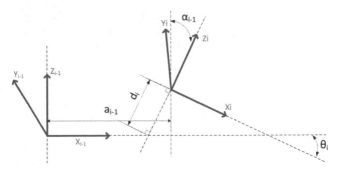

Fig. 8. DH parameters between two coordinate systems

$$T = \begin{bmatrix} & R & & P \\ 0 & 0 & 0 & 1 \end{bmatrix} \qquad (25)$$

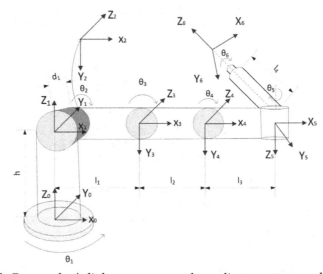

Fig. 9. The nearly Puma robot's link parameters and coordinate systems replacement

I	α_{i-1}	a_{i-1}	d_i	θ_i
1	0	0	h	θ_1
2	$-\pi/2$	0	d_1	θ_2
3	0	l_1	0	θ_3
4	0	l_2	0	θ_4
5	$-\pi/2$	l_3	0	θ_5
6	$\pi/2$	0	l_4	θ_6

Table 1. The DH parameters of the robot

Where, T and R are 4x4 transformation and 3x3 rotation matrices between two consecutive links respectively while P is 3x1 position vector. Detailed information about defining this matrices and vectors can be found in (Kucuk & Bingul, 2006). The transformation matrices of the robot are as follows:

$$
{}^0_1T = \begin{bmatrix} C\theta_1 & -S\theta_1 & 0 & 0 \\ S\theta_1 & C\theta_1 & 0 & 0 \\ 0 & 0 & 1 & h \\ 0 & 0 & 0 & 1 \end{bmatrix} \quad {}^1_2T = \begin{bmatrix} C\theta_2 & -S\theta_2 & 0 & 0 \\ 0 & 0 & 1 & d_1 \\ -S\theta_2 & -C\theta_2 & 0 & 0 \\ 0 & 0 & 0 & 1 \end{bmatrix} \quad {}^2_3T = \begin{bmatrix} C\theta_3 & -S\theta_3 & 0 & l_1 \\ S\theta_3 & C\theta_3 & 0 & 0 \\ 0 & 0 & 1 & 0 \\ 0 & 0 & 0 & 1 \end{bmatrix}
$$

$$
{}^3_4T = \begin{bmatrix} C\theta_4 & -S\theta_4 & 0 & l_2 \\ S\theta_4 & C\theta_4 & 0 & 0 \\ 0 & 0 & 1 & 0 \\ 0 & 0 & 0 & 1 \end{bmatrix} \quad {}^4_5T = \begin{bmatrix} C\theta_5 & -S\theta_5 & 0 & l_3 \\ 0 & 0 & 1 & 0 \\ -S\theta_5 & -C\theta_5 & 0 & 0 \\ 0 & 0 & 0 & 1 \end{bmatrix} \quad {}^5_6T = \begin{bmatrix} C\theta_6 & -S\theta_6 & 0 & 0 \\ 0 & 0 & -1 & -l_4 \\ S\theta_6 & C\theta_6 & 0 & 0 \\ 0 & 0 & 0 & 1 \end{bmatrix}
$$

(26)

In the above matrices $C\theta_i$ and $S\theta_i$ (i=1,2,...,6) indicate $\cos\theta_i$ and $\sin\theta_i$ respectively. The forward kinematics of the robot is the forward multiplication of these matrices given by equation 27. The end effector's position in 3D space is defined with the last column of the result matrix.

$$
{}^0_6T = {}^0_1T\,{}^1_2T\,{}^2_3T\,{}^3_4T\,{}^4_5T\,{}^5_6T = T = \begin{bmatrix} R & P \\ 0\ 0\ 0 & 1 \end{bmatrix}
$$

(27)

In this example, the inverse kinematics of the robot was solved using the most general type of GA and PSO. Before starting to solve the problem with an optimization algorithm, it is needed to define the problem, exactly. The problem presented in this example is finding the robot joint angle values while the robot's parameters and the end effector's position are given. According to problem it can be seen that one individual of the population should have six joint's constraints. They are $\theta_1, \theta_2, \theta_3, \theta_4, \theta_5$ and θ_6. So, an individual can be a 1x6 vector. Each component of the individual is for one of the joint's constraints. The range of the each constraint can be defined separately. However, for the sake of simplicity the ranges of the joints were defined between [-pi, pi]. A sample individual can be seen in Figure 10.

π	-π/4	π/2	0	π/5	0

Fig. 10. A sample individual

The most important part of the solution is the object function, since the fitness value of an individual can be defined with the help of this function. In the example, the object function was defined using the desired position values of the end effector of the robot and the obtained position values from individuals using forward kinematic equations of the robot. The Euler distance between the desired position of the end-effector and the results obtained from an individual can be defined as follows.

Let the desired position of the end effector be $P_d = \begin{bmatrix} x_d & y_d & z_d \end{bmatrix}$ and the obtained position knowledge using an individual be $P_i = \begin{bmatrix} x_i & y_i & z_i \end{bmatrix}$. The Euler distance between these two points in 3D space can be obtained as given in equation 28.

$$d_i = \sqrt{\left(x_d - x_i\right)^2 + \left(y_d - y_i\right)^2 + \left(z_d - z_i\right)^2} \tag{28}$$

The individual that offer the smallest distance is the most convenient candidate of the solution. The object function for the problem can be formulated like in equation 29.

$$O_i = pd_i^{\,2} \tag{29}$$

O_i is object function value, p is penalty constant and d_i is the Euler distance calculated by equation 28, for i'th individual.

The problem was solved firstly using GA. The features of the GA was defined as follows: The gene coding type was determined as real-value coding and the selection type was determined as roulette wheel technique while the mutation and the crossover types were determined as one-point crossover and one-point mutation. The GA starts with defining the needed parameters, like crossover and mutation rates, number of individuals, number of iterations (stopping criteria). Then the initial population is defined randomly in the ranges of the constraints and the object and the fitness values of this population are calculated. In each iteration, the elitism, crossover and mutation operations are performed. The new population is generated after these operations and it masked to comply with the range of the constraints. And finally, the new generation's object and fitness values are calculated and the iterations proceeds until to reach the stopping criteria. In the end of the algorithm, the results are presented. The flowchart of the algorithm can be seen in Figure 11.

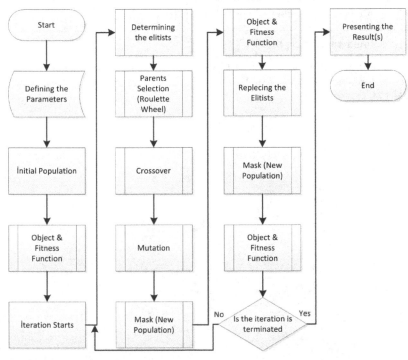

Fig. 11. The flowchart of the GA

The numerical example given below is simply the realization of the algorithm. The robot's first link was replaced in the [0 0 0] coordinates in 3D space and the other link's coordinates were defined according to these coordinates. The aim of the example is to find the robot's joint angles (constraints) that are needed to replace the end-effector of the robot in the goal coordinates. The parameters used in the algorithms were given in Table 2. According to defined parameters, the zero position of the robot's end effector was calculated using zero thetas as [30 15 30]. The solution is the smallest Euler distance between the goal coordinates, [10 5 60], and the each of the individuals. One of the real solutions for this goal coordinates is [0 -pi/2 pi/2 -pi/2 pi/2 0]. The algorithm continues until the stopping criterion is met. In this study, the generation number is the stopping criterion. After running the GA, the constraints' values were found as [-1.0278 -0.5997 -0.6820 -0.8940 0.9335 0] as radian. The coordinates of the end effector can be calculated using these constraints as [9.8143   4.9310 60.0652]. In Figure 12, the two solutions were depicted, the blue one is the sample solution, [0 -pi/2 pi/2 -pi/2 pi/2 0] and the red is the found solution [-1.4238 -1.9043 -0.7396 1.5808 0.9328 0]. The error between the desired and the found end effector's position was [0.1857 0.0690 -0.0652] units.

World Coordinates	Goal Point	Number of Population	Number of Generation	Mutation Rate	Crossover Rate	Penalty	Link Lengths	Zero Position Thetas
[0 0 0]	[10 5 60]	50	2000	0.1	0.9	10	h=30; d1=5; l1=l2=l3=l4=10	[0 0 0 0 0 0]

Table 2. The defined parameters for GA

In this example the most general type of GA was used to solve the problem. It is obvious that more acceptable solutions can be found using different techniques and/or object functions in GA.

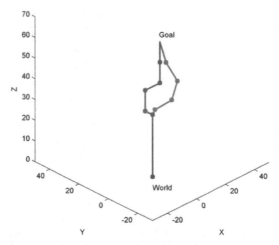

Fig. 12. The sample solution and the solution found with GA

The same problem was solved with PSO algorithm and the solutions were compared with GA. The flowchart of PSO algorithm that was used to solve the problem can be seen in Figure 13 and, the parameters defined for PSO can be seen in Table 3.

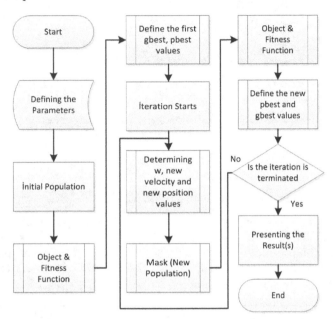

Fig. 13. The flowchart of the PSO algorithm

World Coordinates	Goal Point	Number of Population	Number of Generation	Cognitive Component	Social Component	Penalty	Min. inertia weight	Max. inertia weight	Link Lengths	Zero Position Thetas
[0 0 0]	[10 5 60]	50	2000	2	2	10	0.4	0.4	h=30; d1=5; l1=l2=l3=l4=10	[0 0 0 0 0 0]

Table 3. The defined parameters for PSO

When the algorithm was run, the constraints were found in radian as [-1.3725   -2.5852   1.8546   -0.9442   0.9538   0] and the coordinates of the end effector calculated using these constraints were [9.9979   4.9941   60.0125]. According to these results it can be seen that PSO is found as it serves more convenient solution than GA's solution. In the Figure 14 the two solutions were depicted the blue one is the sample solution, [0 -pi/2 pi/2 -pi/2 pi/2 0] and the red is the found solution [-1.3725 -2.5852 1.8546 -0.9442 0.9538 0].

The convenience of an optimization algorithm to a problem differs from one problem to another and it can only be determined by doing several trials and comparisons. There is no specific algorithm to achieve the best solution for all optimization problems. The comparisons are mainly related to the algorithm's final object function values and the execution time of the algorithm. In the final part of the example a comparison between GA

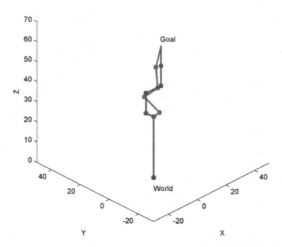

Fig. 14. The sample solution and the found solution using PSO

and PSO for the inverse kinematics problem of the 6 DOF nearly puma robot was presented. For each problem, 100 runs were simulated using the same parameters. The final object function's values and execution times of the two algorithms were presented in a comparative manner using graphs and Tables. The defined parameters for both of the algorithms can be seen in Table 4.

Parameters	GA	PSO
Number of Individuals	100	100
Number of Iterations	1000	1000
Penalty	10	10
World Coordinates	[0 0 0]	[0 0 0]
Goal Coordinates	[10 5 60]	[10 5 60]
Gene Code Type	Real Values	Real Values
Mutation Rate	0.2	***
Crossover Rate	0.9	***
Selection Type	R. Wheel	***
Mutation Type	One Point	***
c1	**	2
c2	**	2
Wmin	**	0.4
Wmax	**	0.9

Table 4. The parameters for GA and PSO for the inverse kinematic problem

Each algorithm was executed under the same condition and on the same computer, and according to the two criterions; the comparison was made between the two algorithms. The first criterion is about the algorithms' final object function values presented in Figure 15. The Figure indicates that both metaheuristic algorithms found the nearly optimum solutions to the problem. On the other hand it can be seen that PSO had more acceptable results than GA. In almost all executions PSO find the exact solutions. However, in 6 executions GA

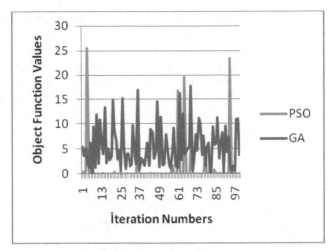

Fig. 15. The object function values of the two algorithms

outperforms PSO. The second criterion is for the execution times of the algorithms. The obtained results, in seconds, were presented in Figure 16. In the Figure, it is obvious that the elapsed time of the PSO is much less than the GA's. As a result, it can be said that PSO produced better results than GA in terms of both final object function's values and the execution times.

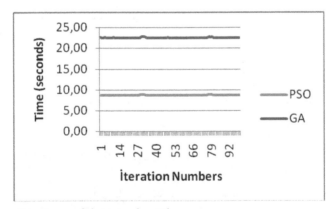

Fig. 16. The execution times of the two algorithms

4.2 Path planning problem for mobile robots

The second example is about path planning for mobile robots. Path planning is one of the most studied topics related to mobile robots. It works for finding the shortest path between the start and goal points while avoiding the collisions with any obstacles. In the example, this problem was solved using GA and PSO.

Path planning can be classified in two categories, global and local path planning. The global type path planning is made for a static and completely known environment while the local path planning is required if the environment is dynamic (Sedighi et al., 2004). In this example, global path planning was performed. The robot's environment was defined as a 10x10 unit size square in 2D coordinate system. In the environment, it was defined several obstacles that have different shapes. The problem is finding the shortest path between pre-defined start and goal points while the coordinates of these points and of the obstacle vertexes are known. A sample environment including obstacles and a sample path can be seen in the Figure 17. In the Figure there are six different shaped obstacles and the path is composed from four points. Two of these points are via points while the others are start and goal points. The mobile robot can reach the goal with tracing the path from the start point. The main object of this problem is to find the shortest path that is not crosses with any obstacles. According to these two constraints, the object function can be defined composing of two parts. The first one is the length of the path and the second one is the penalty function for collisions. The total length of the path can be calculated with the Euler distance between the points on the path if there are two via points that means the path is composed of three parts. The Euler distance between the two points can be simply calculated in 2D environment as in equation 30.

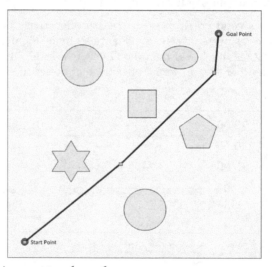

Fig. 17. A sample environment and a path

$$d_i = \sqrt{\left(x_i - x_{i-1}\right)^2 + \left(y_i - y_{i-1}\right)^2} \quad i = 1,2,...,n \tag{30}$$

Where, n is the number of points on the path. The total length of the path is simply the sum of the distances, equation 31. In the equation, L is the total length of the path.

$$L = \sum_{i=1}^{n-1} d_i \quad i = 1,2,...,n \tag{31}$$

The second part of the object function determines the penalty for collision with the obstacles. Determining the collision with an obstacle is quite complicated. The obstacle avoidance technique, used in this study, can be briefly described using Figure 18.

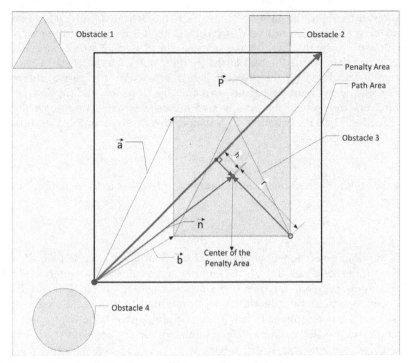

Fig. 18. The penalty calculation for a collision with an obstacle

In the Figure;

Path Area: The rectangular area defined using the maximum and minimum coordinate values of the two points on the path.

Penalty Area: The rectangular area defined using the maximum and minimum coordinate values of the obstacle's vertexes.

$\vec{P}$: The vector between two points of the path, from the first one to the next.

$\vec{a}$ and $\vec{b}$: The vectors from the first point of the $\vec{P}$ to the two vertexes of the penalty area. (There are two additional vectors from the first point of the $\vec{P}$ to the other two vertexes of the penalty area)

$\vec{n}$: The vector from the first point of the $\vec{P}$ to the center of the penalty area.

r : The distance between one vertex and the center of the penalty area.

$h = \dfrac{|\vec{n} \times \vec{P}|}{|\vec{P}|}$: The distance between the center of the penalty area and the $\vec{P}$ vector.

The collision detection procedure starts with defining the path area, and then the obstacles which are totally or partially located in the path are determined like obstacle 2 and obstacle 3 in the Figure 18. After that, penalty areas for each of the determined obstacle are defined; the penalty area for obstacle 3 is shown in the Figure 18. The following processes are achieved for the determined penalty areas. First, the $\vec{P}$ vector between the two points of the path, and four vectors from the first point of the $\vec{P}$ vector to the vertexes of the penalty area are defined. Afterward, the cross products of the $\vec{P}$ with each of these vectors are achieved and the sign of the each product is determined. If the all cross products have the same sign, this means that the part of the path doesn't cross with obstacle. On the other hand if there are different signs obtained from the cross products, then it means that the part of the path crosses with the obstacle. In Figure 18, the signs for the obstacle 2 are all same while not for the obstacle 3 and that means the part of the path crosses with the obstacle 3. Finally the penalty value is calculated for the obstacles which cross with the part of the path. Euler distance between the center of the penalty area and the $\vec{P}$ vector, and the distance between the center and one vertexes of the penalty area are used to calculate the penalty value as defined in equation 32. In the equation, c_p is the penalty constant for the collisions.

$$P = c_p(1 + r - h)^2 \tag{32}$$

As a result, the object function can be formulated completely as in the equation 33.

$$f = L + \sum_{i=1}^{k} P_i \tag{33}$$

Where, **f** is the final object function value, **L** is the total length of the path, **P** is the total penalty value for the collisions and **k** is the number of the collisions for an individual.

The path planning problem for mobile robots was solved using GA and PSO and a comparison between these two algorithms was presented as it is in the first example. The algorithm's flow charts and other details were not given here since they presented in the chapter and also in the first example. The parameters for the problem were defined as in Table 5, and the problem was firstly solved with GA. The GA's parameters were also defined as in Table 6.

Environment	A 10x10 unit area in 2D space
Obstacles	Seven different shape obstacles
Number of Points of an individual	There are four points. Two of them are via points while the others are start and goal points.
Start Point	[0.2, 0.2]
Goal Point	[8.5, 7.5]

Table 5. The parameters for the path planning problem for mobile robots

Number of Individuals	Number of Iterations	Mutation rate	Crossover Rate	Penalty constant
100	200	0.2	0.9	1000

Table 6. GA parameters

The obtained result can be seen in the Figure 19. It can be seen that GA can find nearly optimum solution to the path planning problem for mobile robots.

The second algorithm is the most general type of the PSO like used in first example. The defined parameters for the algorithm are in the Table 7, and the obtained result from PSO is in the Figure 20, and it is obvious that PSO can find nearly optimum solution to the path planning problem for mobile robots.

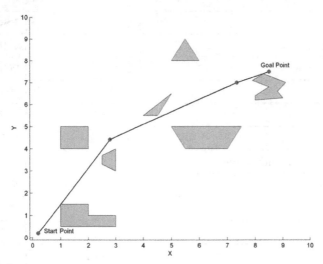

Fig. 19. GA solution

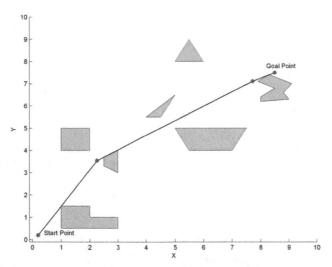

Fig. 20. PSO solution

Number of Individuals	Number of Iterations	Cognitive Component	Social Component	Min. inertia weight	Max. inertia weight	Penalty constant
100	200	2	2	0.4	0.9	1000

Table 7. PSO parameters

For the comparison the same parameters were defined for the two algorithms and each algorithm was run 100 times on the same computer under the same conditions. The parameters were defined as in Table 8.

Parameters	GA	PSO
Number of Individuals	100	100
Number of Iterations	200	200
Penalty	1000	1000
Start Point	[0.2, 0.2]	[0.2, 0.2]
Goal Point	[8.5, 7.5]	[8.5, 7.5]
Number of Points of an individual	3	3
Gene Code Type	Real Values	Real Values
Mutation Rate	0.2	***
Crossover Rate	0.9	***
Selection Type	R. Wheel	***
Mutation Type	One Point	***
c1	**	2
c2	**	2
Wmin	**	0.4
Wmax	**	0.9

Table 8. The parameters for GA and PSO for the path planning problem

The comparison results for the both algorithms were drawn on the two Figures. In Figure 21, there is a graph for the comparison of the final object function values of the algorithms. Lastly, the graph for the comparison of the algorithms' execution times is in the Figure 22.

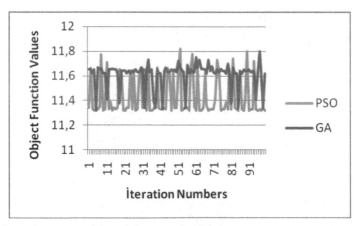

Fig. 21. The object function values of the two algorithms

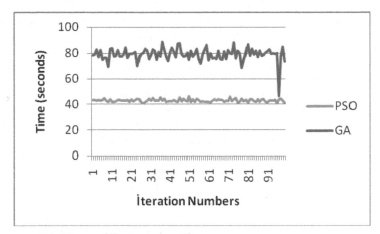

Fig. 22. The execution times of the two algorithms

The perpendicular distance between given start and goal points can be calculated as in equation 34.

$$d = \sqrt{(8.5 - 0.2)^2 + (7.5 - 0.2)^2} = 11.0535 \tag{34}$$

The perpendicular distance was calculated directly from start point to goal point and the obstacles were not considered. If the obstacles were considered that means the distance should be longer. The object function values that were calculated from the algorithms are between 11.3 and 11.82, and this means that these results are nearly optimum results. In terms of the comparison, there is no a significant difference about the final object function values. On the other hand, there is a remarkable difference between the two algorithms' execution times. GA's execution time is nearly twice of the PSO's execution times. So, it can be concluded by saying that the both PSO and GA can be used to solve the path planning of the mobile robots, but if the execution time is important, PSO should be preferred.

5. Conclusion

Heuristic algorithms are an alternative way for the solution of optimization problems. Their implementations are easy. Even if they couldn't find the exact optimum point, they find satisfactory results, in a reasonable solution time. In this chapter, two well-known optimization problems in Robotic were solved with PSO and GA. The first problem is the inverse kinematics of the near PUMA robot while the second problem is the path planning problem for mobile robots. These problems were solved with PSO and GA. It has seen that both algorithm give satisfactory results. But PSO outperforms GA in terms of the solution time, because of the fact that it uses little parameters.

6. References

Barzegar, B.; Rahmani, A.M.; Zamanifar, K. & Divsalar, A. (2009). Gravitational emulation local search algorithm for advanced reservation and scheduling in grid computing systems, *Computer Sciences and Convergence Information Technology,. ICCIT '09. Fourth International Conference on*, Vol., No., pp.1240-1245.

Basturk, B. & Karaboga, D. (2006). An Artificial Bee Colony (ABC) Algorithm for Numeric Function Optimization. *IEEE Swarm Intelligence Symposium 2006*, Vol. 8, No. 1, pp. 687-697.

Blum, C. & Roli, A., Metaheuristics in Combinatorial Optimization: Overview and Conceptual Comparasion, Technical Report, *TR/IRIDIA/2001-13*, October 2001.

Ceylan, O.; Ozdemir, A. & Dag, H.(2010). Gravitational search algorithm for post-outage bus voltage magnitude calculations, *Universities Power Engineering Conference (UPEC), 2010 45th International* , Vol., No., pp.1-6.

Chandra, R. & Rolland, L.(2011). On solving the forward kinematics of 3RPR planar parallel manipulator using hybrid metaheuristics, *Applied Mathematics and Computation*, Vol. 217, No. 22, pp. 8997-9008

Chen, M. W. & Zalzala, A. M. S.(1997). Dynamic modelling and genetic-based trajectory generation for non-holonomic mobile manipulators, Original Research Article *Control Engineering Practice*, Vol.5, No.1, pp. 39-48

Clerc, M.(1999). The swarm and the queen: Towards a deterministic and adaptive particle swarm optimization. *Proc. Congress on Evolutionary Computation.*, pp 1951-1957.

Dorigo, M.; Maniezzo, V.; Colorni, A. (1996). Ant system: optimization by a colony of cooperating agents, Systems, *Man, and Cybernetics*, Part B: Cybernetics, IEEE Transactions on , Vol.26, No.1, pp.29-41

Eberhart, R & Shi, Y.(2001). Particle swarm optimization: Developments, applications and resources, *in Proc. IEEE Congr. Evol. Comput.*, Vol. 1,No., pp. 81–86

Farmer, D.; Packard, N. & Perelson, A. (1986). The immune system, adaptation and machine learning, *Physica D*, Vol. 2, pp. 187–204

Feo, T. A. & Resende, M. G.C. (1995). Greedy randomized adaptive search procedures, *Journal of Global Optimization*, Vol. 6, No. 2, pp. 109–133

Fisher, M.L.; Alexander, H. G. & Rinnooy, K. (1998) The Design, analysis and implementation of heuristics, *Management Science*, Vol. 34., No.3, pp 263-265.

Gao,M. & Jingwen, T.(2007), Path planning for mobile robot based on improved simulated annealing artificial neural network, *icnc, Third International Conference on Natural Computation 2007*, Vol. 3, No., pp.8-12.

Geem, Z.W.; Kim, J.H. & Loganathan, G.V. (2001). A new heuristic optimization algorithm: harmony search, *Simulation*, Vol. 76 No. 2 pp. 60–68

Glover, F. & Kochenberger, Gary A. (2002). *Handbook of Metaheuristics*, Kluwer Academic Publishers, Norwell, MA.

Goldberg, D. E. (1989). *Genetic Algorithms in Search Optimization and Machine Learning*. Addison Wesley. pp. 41. ISBN 0201157675.

Gueaieb, W. & Miah, M.S.(2008). Mobile robot navigation using particle swarm optimization and noisy RFID communication, *Computational Intelligence for Measurement Systems and Applications, 2008. CIMSA 2008.* Vol. , No., pp. 111 - 116

Hong, D.S. & Cho, H.S.(1999). Generation of robotic assembly sequences using a simulated annealing, *Intelligent Robots and Systems*, IROS '99 Proceedings. *1999 IEEE/RSJ International Conference on* , Vol.2, No., pp.1247-1252.

Jourdan, L.; Basseur, M. & Talbi, E.-G.(2009). Hybridizing exact methods and metaheuristics: A taxonomy, *European Journal of Operational Research*, Vol. 199, No. 3, pp. 620-629.

Kalra, P.; Mahapatra, P.B. & Aggarwal, D.K. (2006). An evolutionary approach for solving the multimodal inverse kinematics, *Mechanism and Machine Theory*, Vol. 41, No.10, pp. 1213–1229

Karsli, G. & Tekinalp, O. (2005). Trajectory optimization of advanced launch system, *Recent Advances in Space Technologies*, RAST 2005. *Proceedings of 2nd International Conference on* ,Vol., No., pp. 374- 378

Kennedy, J.(1997). The particle swarm: Social adaptation of knowledge, *in Proc. IEEE Int. Conf. Evol. Comput.*, Vol. ,No., pp. 303–308.

Kennedy, J.; Eberhart, R. (1995). Particle Swarm Optimization. *Proceedings of IEEE International Conference on Neural Networks*, Vol.4, No., IV. pp. 1942–1948

Kirkpatrick, S.; Gelatt, D.C & Vechhi, M.P. (1983) Optimization by simulated annealing, *Science* Vol.220, No.4598, pp. 671–680

Kucuk, Serdar, & Bingul, Zafer. (2005). The Inverse Kinematics Solutions of Fundamental Robot Manipulators with Offset Wrist, Proceedings of the 2005 IEEE International Conference on Mechatronics, Taipei, Taiwan, July 2005

Kucuk, Serdar, & Bingul, Zafer. (2006). Robot Kinematics: Forward and Inverse Kinematics, In: Industrial Robotics: Theory, Modelling and Control, Sam Cubero, pp. (117-148), InTech - Open Access, Retrieved from < http://www.intechopen.com/ articles/show/title/robot_kinematics__forward_and_inverse_kinematics >

Kumar, J.; Xie, S. & Kean, C. A.(2009). Kinematic design optimization of a parallel ankle rehabilitation robot using modified genetic algorithm, *Robotics and Autonomous Systems*, Vol. 57,No. 10, pp 1018-1027

Laguna, M.(1994). A guide to implementing tabu search. *Investigacion Operative*, Vol.4, No.1, pp 5-25

Lazinca, A. (2009). *Particle Swarm Optimization*, InTech, ISBN 978-953-7619-48-0.

Lee, K. Y. & El-Sharkawi, M. A. (2008). *Modern Heuristic Optimization Techniques Theory and Applications to Power Systems*, IEEE Press, 445 Hoes Lane Piscataway, NJ 08854

Li, J. & Xiao, X.(2008). Multi- Swarm and Multi- Best particle swarm optimization algorithm. *Intelligent Control and Automation, 2008, WCICA 2008, 7th World Congress on* , Vol., No., pp.6281-6286.

Lin, CY & Hajela P.(1992). Genetic algorithms in optimization problems with discrete and integer design variables. *Engineering Optimization* ,Vol. 19, No.4 , pp.309–327.

Martı´, R.; Laguna, M. & Glover F. (2006). Principles of scatter search, *European Journal of Operational Research(EJOR)*, Vol. 169, No. 2, pp. 359–372.

Raghavan, M. (1993). The Stewart Platform of General Geometry Has 40 Configurations. *Journal of Mechanical Design*, Vol. 115, pp. 277-282

Rashedi, E.; Nezamabadi-pour, H. & Saryazdi, S.(2009). GSA: a gravitational search algorithm. *Information Science*, Vol. 179, No.13, pp. 2232–2248

Saruhan, H.(2006). Optimum design of rotor-bearing system stability performance comparing an evolutionary algorithm versus a conventional method. *International Journal of Mechanical Sciences*, Vol. 48, No 12. pp 1341-1351

Sedighi , Kamran. H., Ashenayi , Kaveh., Manikas, Theodore. W., Wainwright, Roger L., & Tai, Heng Ming (2004). Autonomous Local Path Planning for a Mobile Robot Using a Genetic Algorithm, The Congress on Evolutionary Computation, Oregon, Portland, June 2004.

Selene, L. C-M.;Castillo, O. & Luis T. A.(2011). Generation of walking periodic motions for a biped robot via genetic algorithms, *Applied Soft Computing*, In Press, Corrected Proof, Available online 20 May 2011.

Storn, R.(1996). On the usage of differential evolution for function optimization, *Fuzzy Information Processing Society*, NAFIPS, Biennial Conference of the North American , vol. , No., pp.519-523

Valle, Y.; Venayagamoorthy, G. K.; Mohagheghi, S.; Hernandez, J.C. & Harley, R. G.(2008). Particle swarm optimization: Basic Concepts,Variants and Applications in Power Systems, *IEEE Transactions On Evolutionary Computation*, Vol. 12, No. 2, pp 171-195.

Wang, T. (2001). *Global Optimization For Constrained Nonlinear Programming*, Doctor of Philosophy in Computer Science in the Graduate College of the University of Illinois at Urbana-Champaign, Urbana, Illinois

Weise, T. (2008). *Global Optimization Algorithms – Theory and Application*, Online as e-book, University of Kassel, Distributed Systems Group, Copyright (c) 2006-2008 Thomas Weise, licensed under GNU, online available http://www.it-weise.de/

Yin, M.; Hu, Y., Yang, F.; Li, X. & Gu, W. (2011). A novel hybrid K-harmonic means and gravitational search algorithm approach for clustering, *Expert Systems with Applications* Vol.38, No. , pp. 9319–9324.

Zargar, A. N. & Javadi, H.(2009). Using particle swarm optimization for robot path planning in dynamic environments with moving obstacles and target, *Third UKSim European Symposium on Computer Modeling and Simulation*, Vol. , No., pp.60-65

6

Data Sensor Fusion for Autonomous Robotics

Özer Çiftçioğlu and Sevil Sariyildiz
Delft University of Technology,
Faculty of Architecture, Delft
The Netherlands

1. Introduction

Multi-sensory information is a generic concept since such information is of concern in all robotic systems where information processing is central. In such systems for the enhancement of the accurate action information redundant sensors are necessary where not only the number of the sensors but also the resolutional information of the sensors can vary due to information with different sampling time from the sensors. The sampling can be regular with a constant sampling rate as well as irregular. Different sensors can have different merits depending on their individual operating conditions and such diverse information can be a valuable gain for accurate as well as reliable autonomous robot manipulation via its dynamics and kinematics. The challenge in this case is the unification of the common information from various sensors in such a way that the resultant information presents enhanced information for desired action. One might note that, such information unification is a challenge in the sense that the common information is in general in different format and different size with different merits. The different qualities may involve different accuracy of sensors due to various random measurement errors. Autonomous robotics constitutes an important branch of robotics and the autonomous robotics research is widely reported in literature, e.g. (Oriolio, Ulivi et al. 1998; Beetz, Arbuckle et al. 2001; Wang and Liu 2004). In this branch of robotics continuous information from the environment is obtained by sensors and real-time processed. The accurate and reliable information driving the robot is essential for a safe navigation the trajectory of which is in general not prescribed in advance. The reliability of this information is to achieve by means of both physical and analytical redundancy of the sensors. The accuracy is obtained by coordinating the sensory information from the redundant sensors in a multi-sensor system. This coordination is carried out by combining information from different sensors for an ultimate measurement outcome and this is generally termed as sensor fusion. Since data is the basic elements of the information, sometimes to emphasize this point the fusion process is articulated with data as *data fusion* where the *sensor fusion* is thought to be as a synonym. Some examples are as follows.

"Data fusion is the process by which data from a multitude of sensors is used to yield an optimal estimate of a specified state vector pertaining to the observed system." (Richardson and Marsh 1988)

"Data fusion deals with the synergistic combination of information made available by various knowledge sources such as sensors, in order to provide a better understanding of a given scene." (Abidi and Gonzales 1992)

"The problem of sensor fusion is the problem of combining multiple measurements from sensors into a single measurement of the sensed object or attribute, called the *parameter.*" (McKendall and Mintz 1992; Hsin and Li 2006)

The ultimate aim of information processing as fusion is to enable the system to estimate the state of the environment and in particular we can refer to the state of a robot's environment in the present case. A similar research dealing with this challenge, namely a multiresolutional filter application for spatial information fusion in robot navigation has been reported earlier (Ciftcioglu 2008) where data fusion is carried out using several data sets obtained from wavelet decomposition and not from individual sensors. In contrast with the earlier work, in the present work, in a multi-sensor environment, fusion of sensory information from different sensors is considered. Sensors generally have different characteristics with different merits. For instance a sensor can have a wide frequency range with relatively poor signal to noise ratio or vice versa; the response time of the sensor determines the frequency range. On the other hand sensors can operate synchronized or non-synchronized manner with respect to their sampling intervals to deliver the measurement outcomes. Such concerns can be categorized as matters of *sensor management* although sensor management is more related to the positioning of the sensors in a measurement system. In the present work data fusion sensor fusion and sensor management issues are commonly are referred to as sensor fusion. The novelty of the research is the enhanced estimation of the spatial sensory information in autonomous robotics by means of multiresolutional levels of information with respect to sampling time intervals of different sensors. Coordination outcome of such redundant information reflects the various merits of these sensors yielding enhanced positioning estimation or estimate the state of the environment. To consider a general case the sensors are operated independently without a common synchronizing sampling command, for instance. The multiresolutional information is obtained from sensors having different resolutions and this multiple information is synergistically combined by means of inverse wavelet transformation developed for this purpose in this work. Although wavelet-based information fusion is used in different applications (Hong 1993; Hsin and Li 2006), its application in robotics is not common in literature. One of the peculiarities of the research is essentially the application of wavelet-based dynamic filtering with the concept of multiresolution as the multiresolution concept is closely tied to the discrete wavelet transform. The multiresolutional dynamic filtering is central to the study together with the Kalman filtering which has desirable features of fusion. Therefore the vector wavelet decomposition is explained in some detail. For the information fusion process extended Kalman filtering is used and it is also explained in some detail emphasizing its central role in the fusion process. In an autonomous robot trajectory the estimation of angular velocity is not a measurable quantity and it has to be estimated from the measurable state variables so that obstacle avoidance problem is taken care of. The angular velocity estimation in real-time is a critical task in autonomous robotics and from this viewpoint, the multiresolutional sensor-based spatial information fusion process by Kalman filtering is particularly desirable for enhanced robot navigation performance. In particular, the multiresolutional sensors provide diversity in the information subject to fusion process. In this way different quality of information with respective merits are synergistically combined.

The motivation of this research is the use of a vision robot for an architectural design and the architectural artifacts therein from the viewpoint of human perception, namely to investigate the perceptual variations in human observation without bias. The similar perception centered research by a human can have an inherent bias due to the interests and background of that human. In this respect, a robot can be viewed as an impartial observer with emulated human perception. Therefore, in this research, the sensory information is treated as robot's visual perception as an emulation of that of a human. A theory for the human perception from a viewpoint of perception quantification and computation is presented earlier (Ciftcioglu, Bittermann et al. 2007; Ciftcioglu 2008). The robot can be a physical real artifact autonomously wandering in an architectural environment. Or alternatively, it can be a virtual robot, wandering in a virtual reality environment. Both cases are equally valid utilization options in the realm of perceptual robotics in architecture. Apart from our interest on human perception of architectural artifacts as motivation, the present research is equally of interest to other adjacent robotics research areas like social robots which are closely related to perception robots. Namely, thanks to the advancements in robotics, today the social robots are more and more penetrating in social life as an aid to many human endeavors. With the advent of rapid progresses in robotics and evolutions on hardware and software systems, many advanced social, service and surveillance mobile robots have been coming into realization in the recent decades; see for instance, *http:/spectrum.ieee.org/robotics*. One of the essential merits of such robots is the ability to detect and track people in the view in real time, for example in a care center. A social robot should be able to keep eye on the persons in the view and keep tracking the persons of concern for probable interaction (Bellotto and Hu 2009). A service robot should be aware of people around and track a person of concern to provide useful services. A surveillance robot can monitor persons in the scene for the identification of probable misbehavior. For such tasks, detecting and tracking multiple persons in often crowded and cluttered scenes in public domain or in a working environment is needed. In all these challenging scenarios perceptual mobile robotics can give substantial contribution for the functionality of such special variety of robots in view of two main aspects. One aspect is vision, which is not the subject-matter of this work. The other aspect is the sensor-data fusion for effective information processing, which is the subject matter of this research where Kalman filtering is the main machinery, as it is a common approach in mobile robotics for optimal information processing.

The further organization of the present work is as follows. After the description of Kalman filtering and wavelet transform in some detail, detailed description of optimal fusion process of information from different multiresolutional levels is presented. The optimality is based on minimum fusion estimation error variance. Finally, autonomous robot implementation is described with the computer experiments the results of which are illustrated by means of both true and estimated trajectories demonstrating the effective multisensor-based, multiresolutional fusion. The work is concluded with a brief discussion and conclusions.

2. Kalman filter

2.1 Description of the system dynamics

Kalman filtering theory and its applications are well treated in literature (Jazwinski 1970; Gelb 1974; Kailath 1981; Maybeck 1982; Brown 1983; Sorenson 1985; Mendel 1987; Grewal and Andrews 2001; Simon 2006). In order to apply Kalman filtering to a robot movement the system dynamics must be described by a set of differential equations which are in state-space form, in general

$$\frac{dx}{dt} = Fx + Gu + w \tag{1}$$

where x is a column vector with the states of the system, F the system dynamics matrix, u is the control vector , and w is a white noise process vector. The process noise matrix Q is related to the process-noise vector according to

$$Q = E[ww^T] \tag{2}$$

The measurements are linearly related to the states according to

$$z = Hx + v \tag{3}$$

where z is the measurement vector, H is the measurement matrix, and v is measurement noise vector which is element-wise white. The measurement noise matrix R is related to the measurement noise vector v according to

$$R = E[vv^T] \tag{4}$$

In discrete form, the Kalman filtering equations become

$$x_k = \Phi_k x_{k-1} + K_k(z_k\text{-}H\Phi_k x_{k\text{-}1}\text{-}HG_k u_{k\text{-}1}) + G_k u_{k\text{-}1}$$
$$z_k = Hx_k + v_k \tag{5}$$

where Φ_k system transition matrix, K_k represents he Kalman gain matrix and G_k is obtained from

$$G_k = \int_0^T \Phi(\tau)G(\tau)d\tau \tag{6}$$

where T is the sampling time interval and the computation of $\Phi(t)$ is given shortly afterwards in (13). In this research information processing from the sensors for estimation is concerned. The control signal (u) is not involved in the filtering operation. Hence the Kalman filtering equation for this case becomes

$$x_k = \Phi_k x_{k-1} + K_k(z_k\text{-}H\Phi_k x_{k\text{-}1}) \tag{7}$$

While the filter is operating, the Kalman gains are computed from the matrix Riccati equations:

$$M_k = \Phi_k P_{k-1} \Phi_k^T + Q_k$$
$$K_k = M_k H^T (HM_k H^T + R_k)^{-1}$$
$$P_k = (I - K_k H)M_k \tag{8}$$

where P_k is a covariance matrix representing errors in the state estimates after an update and M_k is the covariance matrix representing errors in the state estimates before an update. The discrete process noise matrix Q_k can be found from the continuous process-noise matrix Q according to

$$Q_k = \int_0^T \Phi(\tau)Q\Phi^T(\tau)d\tau \tag{9}$$

where $\Phi(t)$ is given in (13). When robot movement is along a straight line with a constant speed v_x, the x component of the system dynamic model is given by

$$x = a_o + v_x t \tag{10}$$

It is to note that the angular speed $\omega=0$. The system dynamics in state-space form is given by

$$\begin{bmatrix} \dot{x} \\ \ddot{x} \end{bmatrix} = \begin{bmatrix} 0 & 1 \\ 0 & 0 \end{bmatrix}\begin{bmatrix} x \\ \dot{x} \end{bmatrix} \tag{11}$$

Where the system dynamics matrix F is given by

$$F = \begin{bmatrix} 0 & 1 \\ 0 & 0 \end{bmatrix} \tag{12}$$

The system transition matrix Φ is computed from inverse Laplace transform of the form

$$\Phi(t) = L^{-1}\left\{ (sI - F)^{-1} \right\} = e^{Ft}$$
$$\Phi_k = \Phi(t) = \begin{bmatrix} 1 & t \\ 0 & 1 \end{bmatrix} \tag{13}$$

The discrete fundamental matrix, i.e., system transition matrix can be found from preceding expression by simply replacing time with the sampling time interval of the perception measurements T or

$$\Phi_k = \Phi(T) = \begin{bmatrix} 1 & T \\ 0 & 1 \end{bmatrix} \tag{14}$$

In two dimensional navigation space, i.e., xy plane, the system transition matrix becomes

$$\Phi_k = \begin{bmatrix} 0 & T & 0 & 0 \\ 0 & 1 & 0 & 0 \\ 0 & 0 & 1 & T \\ 0 & 0 & 0 & 1 \end{bmatrix} \tag{15}$$

and the corresponding state vector X is given by

$$X = [x \quad v_x \quad y \quad v_y] \tag{16}$$

In the case of $\omega \neq 0$, i.e., circular movement with an angular velocity, we consider the geometry shown in Figure 1.

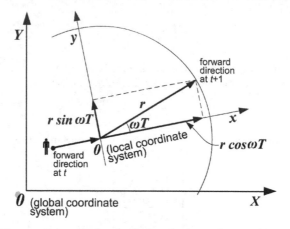

Fig. 1. Geometry with respect to angular deviation during robot navigation

At the local coordinate system, the state variables are

$$x_1 = r\cos(\omega t)$$

$$\dot{x}_1 = -r\omega\sin(\omega t)$$

$$x_2 = r\sin(\omega t)$$

$$\dot{x}_2 = r\omega\cos(\omega t)$$

(17)

So that the system dynamics in state-space form in continuous time is given by

$$\begin{bmatrix} \dot{x}_1 \\ \ddot{x}_1 \\ \dot{x}_2 \\ \ddot{x}_2 \end{bmatrix} = \begin{bmatrix} 0 & 1 & 0 & 0 \\ 0 & 0 & 0 & -\omega \\ 0 & 0 & 0 & 1 \\ 0 & \omega & 0 & 0 \end{bmatrix} \begin{bmatrix} x_1 \\ \dot{x}_1 \\ x_2 \\ \dot{x}_2 \end{bmatrix}$$

(18)

The system transition matrix Φ^o_k is computed from inverse Laplace transform of the form

$$\Phi^\omega_k(t) = L^{-1}\left\{(sI - F_\omega)^{-1}\right\} = e^{F_\omega t}$$

(19)

where $(sI - F_\omega)$ is given by

$$sI - F_\omega = \begin{bmatrix} s & -1 & 0 & 0 \\ 0 & s & 0 & \omega \\ 0 & 0 & s & -1 \\ 0 & -\omega & 0 & s \end{bmatrix}$$

(20)

The inverse of (20) yields

$$
\begin{bmatrix}
\dfrac{1}{s} & \dfrac{1}{s^2+\omega^2} & 0 & -\dfrac{\omega}{s(s^2+\omega^2)} \\[2ex]
0 & \dfrac{1}{s(s^2+\omega^2)} & 0 & -\dfrac{\omega}{s^2+\omega^2} \\[2ex]
0 & \dfrac{\omega}{s(s^2+\omega^2)} & \dfrac{1}{s} & \dfrac{1}{s^2+\omega^2} \\[2ex]
0 & \dfrac{\omega}{s^2+\omega^2} & 0 & \dfrac{1}{s(s^2+\omega^2)}
\end{bmatrix}
\tag{21}
$$

and the inverse Laplace transform of (21) gives the system transition matrix, for $t=T$, as

$$
\Phi_k^\omega =
\begin{bmatrix}
1 & \dfrac{\sin(\omega T)}{\omega} & 0 & \dfrac{\cos(\omega T)-1}{\omega} \\[2ex]
0 & \cos(\omega T) & 0 & -\sin(\omega T) \\[2ex]
0 & -\dfrac{\cos(\omega T)-1}{\omega} & 1 & \dfrac{\sin(\omega T)}{\omega} \\[2ex]
0 & \sin(\omega T) & 0 & \cos(\omega T)
\end{bmatrix}
\tag{22}
$$

During the robot navigation we have obtained two system dynamics models; namely rectilinear straight-ahead and angular rotation cases. To endow the robot to be autonomous the angular velocity should be computed during the navigation. If the perception measurements yield a significant angular velocity, the system dynamics model should switch from linear to non-linear. It is interesting to note that if in (22) $\omega=0$, then it reduces to (15) which is the transition matrix for linear case. In other words, (22) represents inherently the linear case, as well as the rotational robot navigation. If the angular velocity is computed at each step of navigation and if it is non-zero, the robot moves along a non-linear trajectory with each time a deviation θ from linear trajectory. The linear and non-linear cases are illustrated in figure 2 and figure 3 where the measurements are from sensory visual perception (Ciftcioglu, Bittermann et al. 2007; Ciftcioglu 2008).

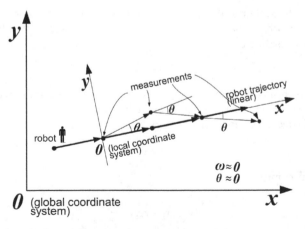

Fig. 2. Measurements along a linear move

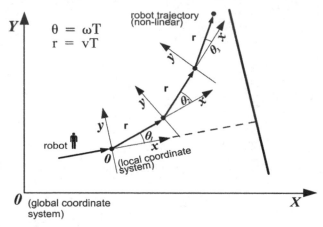

Fig. 3. Robot navigation deviating from a linear move

To compute the angular velocity ω at each step, it is also selected as a state variable, so that the state variables vector given by (16) is modified to

$$X^{\omega} = [x \quad v_x \quad y \quad v_y \quad \omega] \tag{23}$$

However, in this case the system transition matrix in (22) becomes non-linear with respect to ω. In this case Kalman filtering equations should be linearized. This is a form known as extended Kalman filtering.

2.2 Extended Kalman filtering (EKF)

The non-linear state-space form as a set of first-order non-linear differential equations is given by

$$\dot{x} = f(x) + w \tag{24}$$

where x is a vector of the system states, $f(x)$ is a non-linear function of those states, and w is a random zero-mean process. The measurement equation is considered to be a non-linear function of the states according to

$$z = h(x) + v \tag{25}$$

where $h(x)$ is a non-linear measurement matrix, v is a zero-mean random process.
We assume that an *approximate* trajectory x_o is available. This is referred to as the reference trajectory. The actual trajectory x may then be written as

$$x = x_o + \Delta x \tag{26}$$

Hence, (24) and (25) become

$$\dot{x}_o + \dot{\Delta x} = f(x_o + \Delta x) + w$$
$$z = h(x_o + \Delta x) + v \tag{27}$$

The Taylors series expansion yields the linearized model

$$\dot{x}_o + \Delta \dot{x} \approx f(x_o) + \left[\frac{\partial f}{\partial x}\right]_{x=x_o} \Delta x + w$$

$$z = h(x_o) + \left[\frac{\partial h}{\partial x}\right]_{x=x_o} \Delta x + v$$

(28)

where

$$\frac{\partial f}{\partial x}\bigg|_{x=x_o} = \mathbf{F} = \begin{bmatrix} \dfrac{\partial f_1}{\partial x_1} & \dfrac{\partial f_1}{\partial x_2} & \cdots \\[2mm] \dfrac{\partial f_2}{\partial x_1} & \dfrac{\partial f_1}{\partial x_2} & \cdots \\[2mm] \cdot \\ \cdot \end{bmatrix} \qquad \frac{\partial h}{\partial x}\bigg|_{x=x_o} = \mathbf{H} = \begin{bmatrix} \dfrac{\partial h_1}{\partial x_1} & \dfrac{\partial h_1}{\partial x_2} & \cdots \\[2mm] \dfrac{\partial h_2}{\partial x_1} & \dfrac{\partial h_1}{\partial x_2} & \cdots \\[2mm] \cdot \\ \cdot \end{bmatrix}$$

(29)

If the reference trajectory x_o is chosen to satisfy the differential equation

$$\Delta \dot{x} = f(x_o)$$

(30)

In view of (29) and (30), the system dynamics matrix Φ in discrete form for extended Kalman filtering becomes

$$\Phi_k^\omega = \begin{bmatrix} 1 & \dfrac{\sin(\omega T)}{\omega} & 0 & \dfrac{\cos(\omega T)-1}{\omega} & \omega_x \\[3mm] 0 & \cos(\omega T) & 0 & -\sin(\omega T) & \omega_{\dot{x}} \\[3mm] 0 & -\dfrac{\cos(\omega T)-1}{\omega} & 1 & \dfrac{\sin(\omega T)}{\omega} & \omega_y \\[3mm] 0 & \sin(\omega T) & 0 & \cos(\omega T) & \omega_{\dot{y}} \\[3mm] 0 & 0 & 0 & 0 & 1 \end{bmatrix}$$

(31)

Above, ω_x, ω_y, .. are given by

$$\omega_x = \frac{1}{\omega}\left[(T\cos\omega T - \frac{\sin\omega T}{\omega})\dot{x} + (-T\sin\omega T - \frac{\cos\omega T - 1}{\omega})\dot{y}\right]$$

(32)

$$\omega_{\dot{x}} = \left[-\sin\omega T)\dot{x} - \cos\omega T\,\dot{y}\right]T$$

(33)

$$\omega_y = \frac{1}{\omega}\left[(T\sin\omega T - \frac{1-\cos\omega T}{\omega})\dot{x} + (-T\cos\omega T - \frac{\sin\omega T}{\omega})\dot{y}\right]$$

(34)

$$\omega_{\cdot} \Big|_{y} = \Big[\cos \omega T \, \dot{x} - \sin \omega T \, \dot{y} \Big] T \tag{35}$$

In the extended Kalman filter operation, three measurements are considered. Referring to Figure 3 these are

$$\theta = arctg(y \, / \, x) = arctg(x_3 \, / \, x_1) \quad \text{angle}$$
$$r = \sqrt{x^2 + y^2} = \sqrt{x_1^2 + x_3^2} \quad \text{distance} \tag{36}$$
$$v = \sqrt{v_x^2 + v_y^2} = \sqrt{x_2^2 + x_4^2} \quad \text{velocity}$$

From (36), the linearized measurement matrix in terms of state variables becomes

$$\mathbf{H} = \begin{bmatrix} \dfrac{-1}{x_1^2 + x_3^2} & 0 & \dfrac{x_1}{x_1^2 + x_3^2} & 0 & 0 \\[3mm] \dfrac{x_1}{\sqrt{x_1^2 + x_3^2}} & 0 & \dfrac{x_3}{\sqrt{x_1^2 + x_3^2}} & 0 & 0 \\[3mm] 0 & \dfrac{x_2}{\sqrt{x_2^2 + x_4^2}} & 0 & \dfrac{x_4}{\sqrt{x_2^2 + x_4^2}} & 0 \end{bmatrix} \tag{37}$$

Above x_1, x_2, x_3, x_4 are the state variables which are defined as $x_1=x$, $x_2=y$, $x_3=v_x$, and $x_4=v_y$, respectively.

2.3 Estimation
In this work, the Kalman filter is an estimator of states of a dynamic system with a minimal error (innovation) variance and in this sense it is optimal. In order to explain the estimation in detail, the filter equations taking the discrete time point k as reference are briefly given below. A general dynamic system given in a form

$$x(k + 1) = A(k)x(k) + B(k)w(k) \tag{38}$$

$$z(k) = C(k)x(k) + v(k) \tag{39}$$

is terminologically referred to as *state-space*. Above A is the system matrix; B process noise matrix; C is the measurement matrix. Further, $w(k)$ and $v(k)$ are Gaussian process noise and measurement noise respectively with the properties

$$E\{w(k)\} = 0$$
$$E\{w(k)w(l)^T\} = Q(k) \, for \ \ k = l \tag{40}$$
$$= 0 \ otherwise$$

$$E\{v(k)\} = 0$$
$$E\{v(k)v(l)^T\} = R(k) \, for \ \ k = l \tag{41}$$
$$= 0 \ otherwise$$

The estimation in Kalman filter is accomplished recursively which with the notations in the literature became standard matrix equations formulation and it reads

$$x(k+1|k) = A(k)x(k|k) \tag{42}$$

$$P(k+1|k) = A(k)P(k|k)A(k)^T + B(k)Q(k)B(k)^T \tag{43}$$

$$K(k+1) = \frac{P(k+1|k)C(k+1)^T}{C(k+1)P(k+1|k)C(k+1)^T + R(k+1)} \tag{44}$$

So that the updated as the measurements z are available the updated state variables and covariance matrix are

$$x(k+1|k+1) = x(k+1|k) + K(k+1)[z(k+1) - C(k+1)x(k+1|k)]$$
$$z(k+1) - C(k+1)x(k+1|k) = innovation \tag{45}$$

$$P(k+1|k+1) = [I - K(k+1)C(k+1)]P(k+1|k) \tag{46}$$

N-level multiresolutional dynamic system in a vector form can be described by

$$x^{[N]}(k_N + 1) = A^{[N]}(k_N)x^{[N]}(k_N),$$
$$z^{[i]}(k_i) = C^{[i]}(k_i)x^{[i]}(k_i) + v^{[i]}(k_i) \tag{47}$$
$$i = 1,...,N$$

where i=N is the highest resolution level, so that

$$E\left[w^{[N]}(k_N)\right] = 0,$$
$$E\left[w^{[N]}(k_N)w^{[N]}(l_N)^T\right] = Q^{[N]}(k_N), \quad k = l \tag{48}$$
$$= 0 \quad k \neq l$$

Referring to the measurements $z^{[i]}(k_i)$ at different resolution levels, we write

$$E\left[w^{[i]}(k_i)\right] = 0,$$
$$E\left[w^{[i]}(k_i)w^{[i]}(l_i)^T\right] = Q^{[i]}(k_i), \quad k = l \tag{49}$$
$$= 0 \quad k \neq l$$

and

$$E\left[v^{[i]}(k_i)\right] = 0,$$
$$E\left[v^{[i]}(k_i)v^{[i]}(l_i)^T\right] = R^{[i]}(k_i), \quad k = l \tag{50}$$
$$= 0 \quad k \neq l$$

Kalman filter is used to combine the information from the measurements at different resolutional levels and enhance the state estimation rather than to employ single measurement at each time-step.

3. Wavelets

3.1 Multiresolutional decomposition
Wavelet analysis is basically the projection of data onto a set of basis functions in order to separate different scale information. In particular, in the discrete wavelet transform (DWT) data are separated into wavelet detail coefficients (detail-scale information) and approximation coefficients (approximation-scale information) by the projection of the data onto an orthogonal dyadic basis system (Mallat 1989). In the DWT framework, a signal f(x) is decomposed into approximation and detail components to form a multiresolution analysis of the signal as

$$f(x) = \sum_k a_{jo,k}\phi_{jo,k}(x) + \sum_{j=jo}^{jo+J}\sum_k d_{j,k}\psi_{j,k}(x) \quad j_o, j, k \in \mathbb{Z} \tag{51}$$

where $a_{jo,k}$ denote the approximation coefficient at resolution j_o; $d_{j,k}$ denotes the wavelet coefficient at resolution j; $\phi_{jo,k}(x)$ is a scaling function; $\psi_{j,k}(x)$ is a wavelet function at resolution j, and J is the number of decomposition levels. The coefficients are given by

$$\begin{aligned} a_{jo,k} &= \langle f(x), \phi_{jo,k} \rangle \\ d_{j,k} &= \langle f(x), \psi_{j,k} \rangle \quad j_o, j, k \in \mathbb{Z} \end{aligned} \tag{52}$$

Above $\langle . \rangle$ denotes the inner product in the space of square integrable functions $L^2(\angle)$. Specifically, the dyadic DWT assumes the scaling functions have the property of

$$\phi_{jo,k}(x) = 2^{jo/2}\phi(2^{jo}x - k) \tag{53}$$

and the wavelet functions

$$\psi_{j,k}(x) = 2^{j/2}\psi(2^j x - k) \tag{54}$$

The novel feature of wavelets is that they are localized in time and frequency as to signals. This behaviour makes them convenient for the analysis of non-stationary signals. It is an elementary introduction of wavelets by introducing a scaling function, such that

$$\phi(t) = \sqrt{2}\sum_k g_k\phi(2t - k) \tag{55}$$

A counterpart of this function is called mother wavelet function obtained from

$$\psi(t) = \sqrt{2}\sum_k h_k\phi(2t - k) \tag{56}$$

where l_k and h_k are related via the equation

$$h_k = (-1)^k g_{1-k} \tag{57}$$

The coefficients g_k and h_k appear in the quadrature mirror filters used to compute the wavelet transform. $\phi(t)$ and $\psi(t)$ form orthogonal functions which constitute low-pass and high-pass filters respectively which are spaces in $L_2(\Re)$ where inner product of functions is defined with finite energy. The orthogonal spaces satisfy the property

$$\phi_{m,k}(t) \cup \psi(t)_{m,k} = \phi_{m+1,k}(t) \tag{58}$$

where

$$\phi_{m,k}(t) = 2^{m/2} \phi(2^m t - k) \tag{59}$$

$$\psi_{m,k}(t) = 2^{m/2} \psi(2^m t - k) \tag{60}$$

$m=0$ constitutes the coarsest scale. The simplest filter coefficients are known as Haar filter and given by

$$\begin{aligned} h_h &= [h_1 \ h_2] \\ &= \frac{1}{\sqrt{2}}[1 \quad 1] \end{aligned} \tag{61}$$

$$\begin{aligned} g_h &= [g_1 \ g_2] \\ &= \frac{1}{\sqrt{2}}[1 \ -1] \end{aligned} \tag{62}$$

If one sensor is used at the highest resolutional level i.e., i=3 the measurements at different resolution levels can be obtained by the decomposition scheme shown in figure 4.

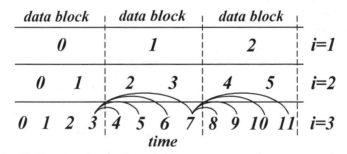

Fig. 4. Measurements at different resolution levels

In this scheme each data block at the highest resolution level (i=3) contains 4 samples. Wavelet decomposition of this block of samples is shown in figure 5.

In figure 5, the measurements are uniform. This means measurement time points in a lower resolution are exactly at the mid of the two points of measurement times at the higher resolutional level, as indicated in figure 4. Within a data block, the state variables at resolution level i are designated as

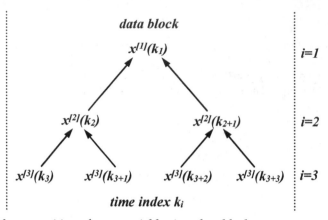

Fig. 5. Wavelet decomposition of state variables in a data block

$$X_m^{[i]} = \begin{bmatrix} x^{[i]}k(i) \\ x^{[i]}k(i+1) \\ .... \\ x^{[i]}k(i+2^{i-1}) \end{bmatrix} \tag{63}$$

$$x^{[i]}{}_{k(i)} = \begin{bmatrix} x_{k,1}^{[i]} \\ x_{k,2}^{[i]} \\ .... \\ x_{k,p}^{[i]} \end{bmatrix} \tag{64}$$

where m is data block index and p is the number of state variables. In a data block, there are 2^{i-1} state variables. Each state variable has p state components. A lower resolution state variable is computed from

$$x^{[i]}(k_i) = h_1 x^{[i+1]}(k_{i+1}) + h_2 x^{[i+1]}(k_{i+1}+1) \tag{65}$$

where h_1 and h_2 are the Haar low-pass filter coefficients. The details component i.e., high frequency part after the decomposition is computed via

$$y^{[i]}(k_i) = g_1 x^{[i+1]}(k_{i+1}) + g_2 x^{[i+1]}(k_{i+1}+1) \tag{66}$$

where g_1 and g_2 are the Haar high-pass filter coefficients. The reconstruction of the states is carried out by combining (65) and (66) in a matrix equation form as given below.

$$\begin{bmatrix} x^{[i+1]}(k_{i+1}) \\ x^{[i+1]}(k_{i+1}+1) \end{bmatrix} = h^{*T} x^{[i]}(k_i) + g^{*T} y^{[i]}(k_i) \tag{67}$$

where h^* and g^* are mirror filters of h and g counterparts; wavelet decomposition and reconstruction is carried out according to the scheme shown in figures 6 and 7.

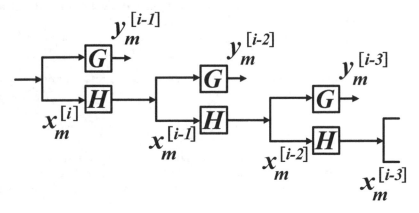

Fig. 6. Wavelet decomposition of state variables in a data block

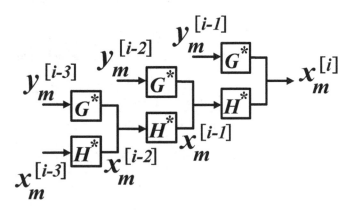

Fig. 7. Wavelet reconstruction of state variables in a data block

The wavelet matrix operator G and the scaling matrix operator H in the decomposition and their counterparts G^* and H^* in the reconstruction contain two-tap Haar filters and they related by

$$G^* = G^T \qquad H^* = H^T \tag{68}$$

The operators G and H are called Quadrature Mirror Filters (QMF) for wavelet decomposition and G^* and H^* QMF for reconstruction. A QMF has the following properties.

$$H^*H + G^*G = 1$$
$$\begin{bmatrix} H^*H & HG^* \\ GH^* & GG^* \end{bmatrix} = \begin{bmatrix} I & 0 \\ 0 & I \end{bmatrix} \tag{69}$$

where I is the identity matrix and (68) implies that filter impulse responses form an orthonormal set. It is to note that the state variable estimations are carried out at the respective resolution levels, as follows.

$$H^{i+1\rightarrow i} = diag\{(1/\sqrt{2})h_h^{(1)},(1/\sqrt{2})h_h^{(2)},.....,(1/\sqrt{2})h_h^{(K)}\} \tag{70}$$

where $h_h^{[i]}$ is scaled two-tap Haar lowpass filters on the diagonal; K is the number of filters involved in a decomposition from the resolution level $i+1$ to i. For instance from 2 to 1, then $i=1$, and K is given by

$$K = 2^{[i-1]} = 1 \tag{71}$$

as this is seen in figure 5 where $2^{[i-1]}$ pairs of state variables in X_m^{i+1} are transformed to $2^{[i-1]}$ lower resolution variables in X_m^i. For the p number of components in a state variable as seen in (64), the H the scaling matrix operator is composed of p number of $H^{i+1\rightarrow i}$ matrices at the diagonal as

$$H = diag\{H_{[1]}^{i+1\rightarrow i},H_{[2]}^{i+1\rightarrow i},......,H_{[p]}^{i+1\rightarrow i}\} \tag{72}$$

where each $H_{[i]}^{i+1\rightarrow i}$ is given by (70). Similarly, for the reconstruction filter, we write

$$G^{i+1\rightarrow i} = diag\{(\sqrt{2})g_h^{(1)},(\sqrt{2})g_h^{(2)},.....,(\sqrt{2})g_h^{(K)}\} \tag{73}$$

The wavelet matrix operator for G for the wavelet coefficients at resolution level i from the resolution level $i+1$

$$G = diag\{G_{[1]}^{i+1\rightarrow i},G_{[2]}^{i+1\rightarrow i},......,G_{[p]}^{i+1\rightarrow i}\} \tag{74}$$

where K is given by (71). For the inverse transform scheme given by figure 7, we write

$$H^* = diag\{H_{[1]}^{i\rightarrow i+1},H_{[2]}^{i\rightarrow i+1},......,H_{[p]}^{i\rightarrow i+1}\} \tag{75}$$

and

$$G^* = diag\{G_{[1]}^{i\rightarrow i+1},G_{[2]}^{i\rightarrow i+1},......,G_{[p]}^{i\rightarrow i+1}\} \tag{76}$$

Where each $H_{[i]}^{i\rightarrow i+1}$ and $G_{[i]}^{i\rightarrow i+1}$ is given by

$$H^{i\rightarrow i+1} = diag\{(\sqrt{2})h_h^{T(1)},(\sqrt{2})h_h^{T(2)},.....,(\sqrt{2})h_h^{T(K)}\} \tag{77}$$

$$G^{i\rightarrow i+1} = diag\{(1/\sqrt{2})g_h^{T(1)},(1/\sqrt{2})g_h^{T(2)},.....,(1/\sqrt{2})g_h^{T(K)}\} \tag{78}$$

Above T indicates transpose.

3.2 Multiresolution by sensory measurements
In subsection A wavelet decomposition is presented where N-level wavelet decomposition scheme lower level measurements are obtained basically by means of wavelet decomposition. This implies that for a state variable all measurements are obtained by a single sensor associated with that state variable. However in the present case we consider

multiple sensors for the same state variable while the sensors are operated in different resolutions. Referring to figure 4, 3 sensors of different resolutions are considered. Since sensors are operated independently the measurements are non-uniform, in general. This means measurements in the lower resolution are not necessarily at the mid of the two points of the measurement times at the higher resolutional level. This is depicted in figure 8. Uniform sampling is seen in figure 4.

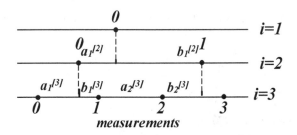

Fig. 8. Non-uniform sampling in a data block of three resolutions

Before explaining the fusion process, the wavelet decomposition for the non-uniform case will be explained in detail since this is central to this study. Decomposing the state variables at time indices 0 and 1 at resolution level $i=3$ into a single state variable at time index 0 at resolution level $i=2$ can be achieved by lowpass filter $h^{[2]}(1)$ as follows.

$$h^{[2]}(1) = \left[\frac{b_1^{[3]}}{a_1^{[3]} + b_1^{[3]}} \quad \frac{a_1^{[3]}}{a_1^{[3]} + b_1^{[3]}} \right] \tag{79}$$

which can be written in general form

$$h^{[i]}(1) = \left[\frac{b_{k_i}^{[i+1]}}{a_{k_i}^{[i+1]} + b_{k_i}^{[i+1]}} \quad \frac{a_1^{[i+1]}}{a_{k_i}^{[i+1]} + b_{k_i}^{[i+1]}} \right] \tag{80}$$

where index i denotes the resolution level; k_i is the time index at the resolution level i; $a_{ki}^{[i+1]}$ and $b_{ki}^{[i+1]}$ are the relative time intervals. The lowpass filter $h^{[i]}(k_i)$ for deriving a coarsened estimate state variable at time index k_i and at resolution level i is based on the appropriate pair of estimated state variables at resolution level $i+1$. As the lowpass filter is determined, the highpass filter and the inverse filters can be determined by the filterbank implementation of the Quadrature Mirror Filter (QMF) shown in figures 6 and 7. Hence from lowpass filter $h^{[i]}(k_i)$ the highpass filter $g^{[i]}k(i)$ and the inverse filters $h_{inv}^{[i]}(k_i)$ and $g_{inv}^{[i]}(k_i)$ are determined as given below that they satisfy the constraints given by (69).

$$g^{[i]}(k_i) = \left[\frac{2b_{k_i}^{[i+1]}}{a_{k_i}^{[i+1]} + b_{k_i}^{[i+1]}} \quad \frac{-2a_1^{[i+1]}}{a_{k_i}^{[i+1]} + b_{k_i}^{[i+1]}} \right] \tag{81}$$

$$h_{inv}^{[i]}(k_i) = \left[\frac{0.5(a_{k_i}^{[i+1]} + b_{k_i}^{[i+1]})}{b_{k_i}^{[i+1]}} \quad \frac{0.5(a_{k_i}^{[i+1]} + b_{k_i}^{[i+1]})}{a_{k_i}^{[i+1]}} \right] \tag{82}$$

and

$$g_{inv}^{[i]}(k_i) = \begin{bmatrix} \dfrac{0.5(a_{k_i}^{[i+1]} + b_{k_i}^{[i+1]})}{2b_{k_i}^{[i+1]}} & \dfrac{-0.5(a_{k_i}^{[i+1]} + b_{k_i}^{[i+1]})}{2a_{k_i}^{[i+1]}} \end{bmatrix} \tag{83}$$

For $a_{ki}^{[i+1]}=b_{ki}^{[i+1]}$, the filters reduce to Haar filters given (61) and (62) and shown in figures 6 and 7.

In this implementation the scaling and wavelet operators H and G for decomposition i+1→i are given by

$$H^{i+1 \to i} = diag\left\{ h^{[i]}(k_i), h^{[i]}(k_i + 1), ..., h^{[i]}(k_i + 2^{i-1} - 1) \right\}$$
$$G^{i+1 \to i} = diag\left\{ g^{[i]}(k_i), g^{[i]}(k_i + 1), ..., g^{[i]}(k_i + 2^{i-1} - 1) \right\} \tag{84}$$

For $a_{ki}^{[i+1]}= b_{ki}^{[i+1]}$ (84) reduces to (70) and (73).

The inverse scaling and wavelet operators H and G for construction i→i+1 are given by

$$H^{i \to i+1} = diag\left\{ h_{inv}^{[i]}(k_i)^T, h_{inv}^{[i]}(k_i + 1)^T, ..., h_{inv}^{[i]}(k_i + 2^{i-1} - 1)^T \right\}$$
$$G^{i \to i+1} = diag\left\{ g_{inv}^{[i]}(k_i)^T, g_{inv}^{[i]}(k_i + 1)^T, ..., g_{inv}^{[i]}(k_i + 2^{i-1} - 1)^T \right\} \tag{85}$$

For $a_{ki}^{[i+1]}= b_{ki}^{[i+1]}$ (84) and (85) reduces to (77) and (78).

4. Fusion process as multiresolutinal dynamic filtering (MDF)

The fusion of information is central to this research. Therefore, in the preceding section wavelet decomposition and reconstruction is presented in vector form for the sake of explaining the fusion process in detail. However the wavelet decomposition in this work is not used. This is simply because lower resolution level sensory measurements are obtained from associated sensors and not from wavelet decomposition of the highest resolution level sensory measurements. Therefore only the wavelet reconstruction is relevant. The lower resolution level measurements are used to update the estimated information at this very level. Afterwards, this information is transformed to higher resolution level information by inverse wavelet transform where the inverse transformation wavelet coefficients, that is the detail coefficients, are not involved in this process as they are all zero. Because of this reason the transformed information at a higher resolution level is the same as the information lying in the preceding lower level. But this very information at the higher resolution level timely coincides with the sensory information of this level. This is achieved by non-uniform formulation of wavelet transform. By doing so, the independent operation of multiresolutional sensors is aimed to make the information fusion effective. The actual implementation in this work is explicitly as follows. Referring to figure 8, a data block has four sensory measurement samples at the highest resolution (i=3) and one sensory sample in the lowest resolution (i=1). The resolution level between the highest and lowest contains two sensory measurements. By means of inverse wavelet transform the updated estimations at levels i=1 and i=2 are transformed to highest level separately providing the estimate of the signal of the resolution index i=1 and i=2 and the highest level (i=3). In the level one, a single estimation, in the level two, two updated estimations are projected to highest level. In the level three the estimations are updated for four samples. At all levels the estimations are updated by Kalman filtering for $i=1,2$, and 3. Signals from different resolutional levels are projected to the highest resolution level so that they all have four samples in a data block. The basic update scheme for dynamic multiresolutional filtering is shown in Fig. 9 where at each resolutional level, when the measurement Z is available, the state variables are updated and when the block m is complete the inverse wavelet transform and fusion is performed. During the inverse transformation the wavelet coefficients are all zero due to non-performed wavelet decomposition.

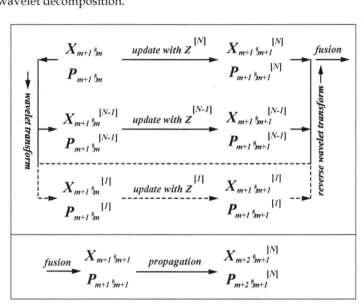

Fig. 9. Wavelet decomposition of state variables in a data block m

Explicitly, the basic update scheme is as follows.

$$X_{m+1|m+1}{}^{[i]} = X_{m+1|m}{}^{[i]} + K_{m+1}{}^{[i]}(Z_{m+1}{}^{[i]} - C_{m+1}{}^{[i]}X_{m+1|m}{}^{[i]}) \tag{86}$$

and

$$P_{XX_{m+1|m+1}}{}^{[i]} = (I - K_{m+1}{}^{[i]}C_{m+1}{}^{[i]})P_{XX_{m+1|m}}{}^{[i]} \tag{87}$$

The minimum variance Kalman gain matrix $K_{m+1}{}^{[i]}$ at each level, is determined by

$$K_{m+1}{}^{[i]} = P_{XX_{m+1|m}}{}^{[i]}C_{m+1}{}^{[i]T}\left(C_{m+1}{}^{[i]}P_{XX_{m+1|m}}{}^{[i]}C_{m+1}{}^{[i]T} + R_{m+1}{}^{[i]}\right)^{-1} \tag{88}$$

where the measurent matrix $C_{m=1}{}^{[i]}$ and $R_{m+1}{}^{[i]}$ are given by

$$C^{[i]}{}_{m+1} = diag\begin{bmatrix} C^{[i]}[(m+1)2^{i-1}], C^{[i]}[(m+1)2^{i-1} - 2^{i-1} + 1], ..., \\ ..... \qquad , C^{[i]}[(m+1)+2^{i-1}+1] \end{bmatrix} \tag{89}$$

$$R_{m+1}^{[i]} = diag \begin{bmatrix} R^{[i]}[(m+1)2^{i-1}], R^{[i]}[(m+1)2^{i-1} - 2^{i-1} + 1], ..., \\ ..... \qquad\qquad , R^{[i]}[(m+1) + 2^{i-1} + 1] \end{bmatrix} \qquad (90)$$

Once, the sequences of updated state variables and error covariances $X_{m+1|m+1}^{[N,i]}$ and $P_{m+1|m+1}^{[N,i]}$ for $i=1,2,..,N$, are determined, they must be fused to generate an optimal $X_{m+1|m+1}^{[NF]}$ and $P_{m+1|m+1}^{[NF]}$. For the minimum fusion error covariance $P_{m+1|m+1}^{[NF]}$ as derived in (Hong 1991; Hong 1992), the fused estimate $X_{m+1|m+1}^{[NF]}$ is calculated as

$$X_{m+1|m+1}^{[NF]}$$
$$= P_{m+1|m+1}^{[NF]} \left[\sum_{i=1}^{N} \left(P_{m+1|m+1}^{[N,i]} \right)^{-1} X_{m+1|m+1}^{[N,i]} - (N-1) \left(P_{m+1|m}^{[N]} \right)^{-1} X_{m+1|m}^{[N]} \right] \qquad (91)$$

where the minimum fusion error covariance $P_{m+1|m+1}^{[NF]}$ is given by

$$\left(P_{m+1|m+1}^{[NF]} \right)^{-1} = \sum_{i=1}^{N} \left(P_{m+1|m+1}^{[N,i]} \right)^{-1} - (N-1) \left(P_{m+1|m}^{[N]} \right)^{-1}. \qquad (92)$$

The fused estimate $X_{m+1|m+1}^{[NF]}$ is a weighted summation of both predicted $X_{m+1|m}^{[N]}$ and updated $X_{m+1|m+1}^{[N,i]}$, for $i=1,2,..,N$. The sum of the weight factors equal to the identity I. This can be seen by substitution of $P_{m+1|m+1}^{[NF]}$ given above into the expression of $X_{m+1|m+1}^{[NF]}$ in (91). With the estimations in different level of resolutions and finally fusion of the level-wise estimations for unified estimations form a multiresolutional distributed filtering (MDR).

5. Experiments with the autonomous robot

The computer experiments have been carried out with the simulated robot navigation. The state variables vector is given by (93) where $N=i$ to represent a general resolutional level.

$$x^{[N]}{}_k(N) = \begin{bmatrix} x_{k,1}^{[N]} \\ x_{k,2}^{[N]} \\ .... \\ x_{k,p}^{[N]} \end{bmatrix} \qquad (93)$$

Explicitly,

$$x^{[N]}k(N) = [x, \dot{x}, y, \dot{y}, \omega]$$

where ω is the angular rate and it is estimated during the move. When the robot moves in a straight line, the angular rate becomes zero and the other state variables namely, x and y coordinates and the respective velocities remain subject to estimation. The overall robot trajectory is shown in figure 10 where there are four lines plotted but they are all close to each other due to the scale involved. Broadly one can see approximately a linear trajectory followed by a curve trajectory and approximately another linear trajectory afterwards. The line marked by * sign represents the measurement of the data at the highest resolution level for $i=3$. The line marked by • is the estimation by sensor fusion. The line marked by + sign is the estimation by extended Kalman filtering for the data obtained from the sensor of the highest resolution level ($i=3$). The line indicated by o sign is the reference trajectory. These lines are not explicitly seen in this figure. For explicit illustration of the experimental outcomes the same figure with different zooming ranges and the zooming powers are given in figures 11-18.

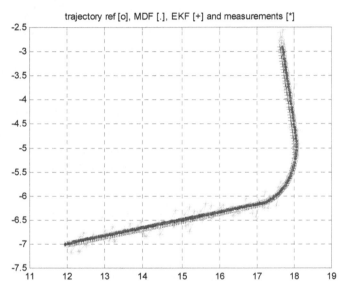

Fig. 10. The overall *measurement, reference, extended Kalman filtering(EKF)* and *multiresolutional distributed filtering* (*MDF*) estimation of robot trajectory. The * sign is for measurement; + for EKF; o for reference; • for *MDF* estimated trajectory.

Figure 11 and 12 shows the estimations in a linear mode. Figure 12 is the enlarged form of figure 11. From these figures it is seen that, the Kalman filtering is effective at the first linear part of the trajectory; namely relative to Kalman filtering estimation, the estimation by sensor fusion by MDF is inferior. In this mode the angular velocity is zero, the system matrix is linear and the linear Kalman filter is accurate enough to describe the dynamic system. During this period, the Kalman filter estimations are carried in smallest sampling time intervals. At the same period MDF estimations made in lower resolution levels are extended to the highest resolution level. However during this extension the x and y coordinates do not match exactly the estimates in the highest resolutional level because of time difference between the estimations. Explicitly, in figure 8 the estimation in the level $i=0$ is extended to estimation number 3 in the resolutional level $i=3$ where there is time difference of more than one sampling time interval. The result is higher estimation error in the MDF and this error appears to be as systematic error in estimation in the form of hangoff error, i.e., error does

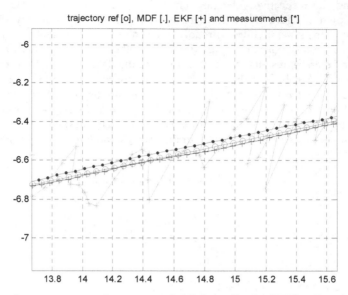

Fig. 11. Enlarged *measurement, reference, extended Kalman filtering(EKF)* and *multiresolutional distributed filtering* (*MDF*) estimation of robot trajectory in first linear period. The * sign is for measurement; + for EKF; o for reference; • for *MDF* estimated trajectory.

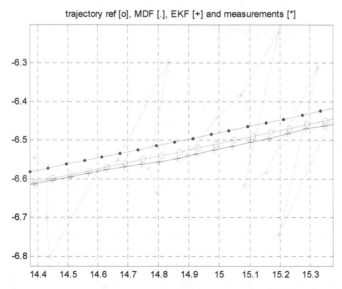

Fig. 12. Enlarged *measurement, reference, extended Kalman filtering(EKF)* and *multiresolutional distributed filtering* (*MDF*) estimation of robot trajectory in first linear period. The * sign is for measurement; + for EKF; o for reference; • for *MDF* estimated trajectory.

not go to zero. In a sensor fusion process such sensor delays are inevitable and thus delay-related effects are inevitable. This research clearly illustrates the extent of such effects which are to be categorized eventually as errors. Consequently one can conclude that the fusion of sensors from different resolutional levels has an inherent peculiarity of latency that turns out to be undesirable outcome in some cases, as it is the case in the present situation, although this is not general as the following figures (13-18) indicate.

Figure 13 and 14 shows the estimations in a bending mode. Figure 14 is the enlarged form of figure 13. In this case estimations by sensor fusion are superior to the estimations by extended Kalman filtering. This can be explained seeing that system matrix involves the angular velocity which makes the system dynamics matrix non-linear. This results in marked separation between the reference trajectory and the estimated trajectory due to the approximation error caused by the Taylor's series expansion and ensuing linearization in the extended Kalman filtering (EKF) in the highest resolution level. One should note that the effect of this approximation error propagated four times in a data block to the time point where fusion and predictions are made for the following data block as seen in figure 4. In this nonlinear period, sensor fusion is very effective and the difference between the estimated outcomes and the reference trajectory is apparently negligibly small. However, this is not exactly so. Because of the delay of the data from the lower resolutional levels to the highest resolutional level as described in the preceding paragraph, there is some difference between the true position and the estimated position. Nevertheless, the true trajectory is almost perfectly identified. The reason for the effectiveness in the lower resolution levels is due to more effective linearization and therefore better state estimations. Although in the lower resolutions levels error in the linearization process for EKF is greater relative to that occurred in the higher resolutional level, such modeling errors are accounted

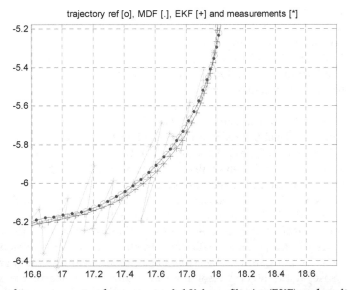

Fig. 13. Enlarged *measurement, reference, extended Kalman filtering(EKF)* and *multiresolutional distributed filtering* (*MDF*) estimation of robot trajectory in the bending period. The * sign is for measurement; + for EKF; o for reference; • for *MDF* estimated trajectory.

trajectory ref [o], MDF [.], EKF [+] and measurements [*]

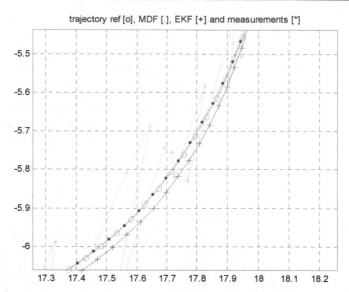

Fig. 14. Enlarged *measurement, reference, extended Kalman filtering(EKF)* and *multiresolutional distributed filtering* (*MDF*) estimation of robot trajectory in the bending period. The * sign is for measurement; + for EKF; o for reference; • for *MDF* estimated trajectory.

for in the process noise and therefore effect of errors due to linearization becomes less important. However, it should be pointed out that, in all resolutional levels, the sensor quality plays essential role on the estimation errors.

Figure 15 shows the trajectory where EKF estimation crosses the reference trajectory. This can be explained as follows. EKF estimations in the highest resolutional level without multiresolutional information, start to deviate from the reference trajectory in the bending mode as seen in figures 13 and 14. In this case Kalman filter tend to make estimations to compensate this deviation error and therefore the deviation start to become smaller. At the same time the bending information namely the angular frequency (ω) becomes effective and these two corrective joint actions in this turbulent transition period make the estimation error minimal and finally the estimated trajectory cross the reference trajectory. It is to note that in this resolutional level the Kalman filter bandwidth is relatively wide justifying the sampling rate which is the highest. As seen in (31), the system matrix is highly involved with the angular frequency and even small estimation error on ω might cause relatively high effects in this non-linear environment. After crossing, the deviations start to increase and after the bending is over it remains constant in the second linear mode in the trajectory, as seen in figures 16 and 17. On the other hand, during this period, the multiresolutional distributed filtering (MDF) estimations improve due to due to incoming bending information, the deviations become smaller and finally it crosses the reference trajectory. This crossing is shown in figure 16. The estimations at the lower resolution level are much accurate than those at the highest resolution level and by means of the sensor fusion process, the fused estimations are quite accurate at the bending mode and afterwards. This is seen in figures 13 through18. Also the effect of the information latency on the position estimation is clearly observed in these figures.

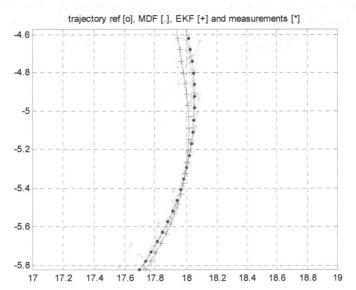

Fig. 15. Enlarged *measurement, reference, extended Kalman filtering(EKF)* and *multiresolutional distributed filtering (MDF)* estimation of robot trajectory in the bending period. The * sign is for measurement; + for EKF; o for reference; • for *MDF* estimated trajectory.

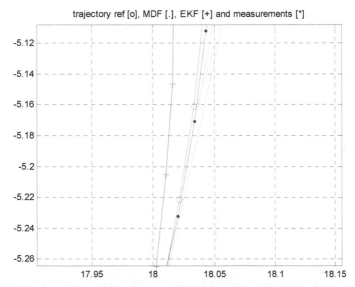

Fig. 16. Enlarged *measurement, reference, extended Kalman filtering(EKF)* and *multiresolutional distributed filtering (MDF)* estimation of robot trajectory in the transition between bending and second linear periods. The * sign is for measurement; + for EKF; o for reference; • for *MDF* estimated trajectory.

Figure 17 and 18 shows the estimations in the second linear trajectory. They are the enlarged form of figure 10 at this very period. Next to satisfactory MDF estimations, the figures show the

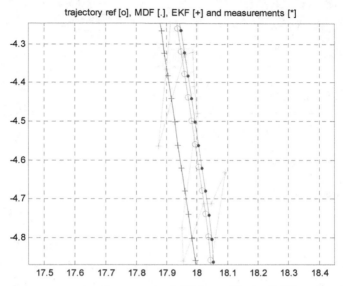

Fig. 17. Enlarged *measurement, reference, extended Kalman filtering(EKF)* and *multiresolutional distributed filtering* (*MDF*) estimation of robot trajectory in the second linear period. The * sign is for measurement; + for EKF; o for reference; • for *MDF* estimated trajectory.

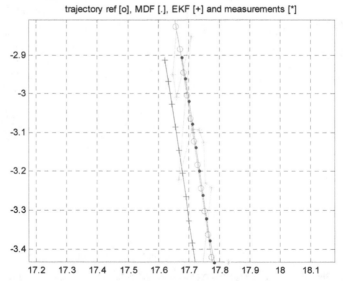

Fig. 18. Enlarged *measurement, reference, extended Kalman filtering(EKF)* and *multiresolutional distributed filtering* (*MDF*) estimation of robot trajectory in the second linear period. The * sign is for measurement; + for EKF; o for reference; • for *MDF* estimated trajectory.

estimations of the Extended Kalman filtering in the highest resolutional level. These estimations have relatively large errors which are due to the system dynamics matrix given by (31) where ω is expected to be zero in this linear period; however it is approximately zero ($\omega \approx 0$) as calculated by Kalman filtering. The small nonzero terms in the matrix cause error in estimations which are interpreted as model errors in the Kalman filtering operation. Further such errors also cause round off errors in the covariance matrix in (43) and finally poor estimation. When the same operation is repeated by switching the matrix forcefully from bending mode to linear mode by putting $\omega=0$ in (31), the Kalman filtering estimation in the last linear period becomes comparable as illustrated in figure 11 and 12. Since the highest resolution level estimations have large errors, they have small contributions to the sensor-data fusion process and therefore the fusion results remain accurate. Figures 13 through 18 represent a firm indication of the effectiveness and robustness of the sensor fusion, in this research.

6. Discussion

In this work the effectiveness of multisensor-based multiresolutional fusion is investigated by means of estimation errors of mobile robot position determination. The comparison is made offline but not real-time. By doing so, a clear view presented about at what conditions the multiresolutional multi-sensor fusion process is effective and also in which circumstances the fusion process may have shortcomings and why. However, the implementation can be carried on in real-time in the form of one block ahead prediction forming the data-sensor fusion, and one step-ahead prediction at the highest resolutional level i.e., for $i=3$ without fusion process. These are illustrated in figure 4. In both cases, i.e., real-time and off-line operations, the merits of the multiresolutional multisensor fusion remains robust although some unfavorable deviation from the existing results in real-time may occur due to a block prediction compared to 1-step-ahead prediction, obviously. Investigations on real-time operation for the assessment of the robustness are interesting since the mobile robot is especially meant for this type of operation.

7. Conclusions

Autonomous mobile robot navigation is a challenging issue where robot should be provided with accurate and reliable position information. Although reliable information can be provided by adding redundant sensors, the enhanced accuracy and precision information can be provided by synergistically coordinating the information from these sensors. In this respect, the present research introduced a novel information fusion concept by inverse wavelet transform using independent multiresolutional sensors. In the linear system description, the highest resolutional sensor provides enough information for optimum information processing by Kalman filtering where residual variance is minimum so that the information delivered by multiresolutional sensors can be redundant depending on the sensors's qualities and associated noises. In situations where system dynamics is non-linear, Kalman filter is still optimal in its extended formulation. However, the estimation errors in this case are dependent on the degree of the non-linearity of the system dynamics. The multiresolutional sensor fusion becomes quite effective in the non-linear case since the partial nonlinearity information of the system in different resolutional scales is available. Sensor quality is always an important factor playing role on the estimation. These features are demonstrated and the fusion process presented can easily be extended to consider real-time operation as well as some cases of probabilistic nature such as missing measurements, sensor failures and other probabilistic occurrences.

8. References

Abidi, M. A. and R. C. Gonzales (1992). Multi-sensor Image Fusion. *Data Fusion in Robotics and Machine Intelligence*. R. S. Blum and Z. Liu, Academic Press.

Beetz, M., T. Arbuckle, et al. (2001). Integrated, Plan-based Control of Autonomous Robots in Human Environments. *IEEE Intelligent Systems* 16(5): 56-65.

Bellotto, N. and H. Hu (2009). Multisensor-based Human Detection and Tracking for Mobile Service Robots. *IEEE Trans. System, Man, and Cybernetics-B* 39(1): 167-181.

Brown, R. G. (1983). *Introduction to Random Signal Analysis and Kalman Filtering*. New York, John Wiley & Sons.

Ciftcioglu, Ö. (2008). Multiresolutional Filter Application for Spatial Information Fusion in Robot Navigation. In: *Advances in Robotics, Automation and Control*, 355-372, J. Aramburo and A. R. Trevino (Eds.), ISBN 78-953-7619-16-9, I-Tech Publishing, Vienna, Austria.

Ciftcioglu, Ö. (2008). Shaping the Perceptual Robot Vision and Multiresolutional Kalman Filtering Implementation. *Int. Journal Factory Automation, Robotics and Soft Computing(3)*: 62-75, ISSN 1828-6984, International Spociety for Advanced Research, *www.internationalsar.org*.

Ciftcioglu, Ö., M. S. Bittermann, et al. (2007). Visual Perception Theory Underlying Perceptual Navigation. *Emerging Technologies, Robotics and Control Systems* International Society for Advanced Research: 139-153.

Gelb, A. (1974). *Applied Optrimal Estimation*. Cambridge, MA, MIT Press.

Grewal, M. S. and A. P. Andrews (2001). *Kalman Filtering Theory and Practice Using MATLAB*. New York, Wiley.

Hong, L. (1991). Adaptive Distributed Filtering in Multi-coordinated Systems. *IEEE Trans. on Aerospace and Electronic Systems* 27(4): 10.

Hong, L. (1992). Distributed Filtering Using Set Models. *IEEE Trans. on Aerospace and Electronic Systems* 28(4): 10.

Hong, L. (1993). Multiresolutional Filtering using Wavelet Transform. *IEEE Trans. on Aerospace and Electronic Systems* 29(4): 1244-1251.

Hsin, H. C. and A. C. Li (2006). Wavelet-based Kalman Filtering in Scale Space for Image Fusion. *Pattern Recognition and Computer Vision*. C. H. C. a. P. S. P. Wang. Singapore, World Scientific.

Jazwinski, A. H. (1970). *Stochastic Processes and Filtering Theory*. New York Academic Press.

Kailath, T. (1981). *Lectures on Wiener and Kalman Filtering*. New York, Springer Verlag.

Mallat, S. G. (1989). A theory for multiresolution signal decomposition:the wavelet representation. *IEEE Trans. on Pattern Analysis and Machine Intelligence*. 11(7): 674-693.

Maybeck, P. S. (1982). *Stochastic Models, Estimation and Control*, Vol II. New York, Academic Press.

McKendall, R. and M. Mintz (1992). Data Fusion Techniques using Robust Statistics. *Data Fusion in Robotics and Machine Intelligence* Academic Press: 211-244.

Mendel, J. M. (1987). *Kalman Filtering and Digital Estimation Techniques*. New York, IEEE.

Oriolio, G., G. Ulivi, et al. (1998). Real-time Map Building and Navigation for autonomous robots in unknown environments. *IEEE Trans. on Systems, Man and Cybernetics - Part B*: Cybernetics 28(3): 316-333.

Richardson, J. M. and K. A. Marsh (1988). Fusion of Multi-Sensor Data. *Int. J. Robotics Research* 7(6): 78-96.

Simon, D. (2006). Optimal State Estimation. New Jersey, Wiley Interscience.

Sorenson, H. W. (1985). *Kalman Filtering: Theory and Application*. New York, IEEE Press.

Wang, M. and J. N. K. Liu (2004). Online Path Searching for Autonomous Robot Navigation. *IEEE Conf. on Robotics, Automation and Mechatronics*, 1-3 December, Singapore: 746-751, vol.2, ISBN 0-7803-8645-0

7

Multi-Criteria Optimal Path Planning of Flexible Robots

Rogério Rodrigues dos Santos[1], Valder Steffen Jr.[1]
and Sezimária de Fátima Pereira Saramago[2]
[1]School of Mechanical Engineering,
Federal University of Uberlândia, Uberlândia, MG
[2]Faculty of Mathematics,
Federal University of Uberlândia, Uberlândia, MG
Brazil

1. Introduction

Determining the trajectory from the initial to the final end-effector positioning represents one of the most common problems in the path-planning design of serial robot manipulators. The movement is established through the specification of a set of intermediate points. In this way, the manipulator is guided along the trajectory without any concern regarding the intermediate configurations along the path. However, there are applications in which the intermediate points have to be taken into account both for path-planning and control purposes. An example of such an application is the case of robot manipulators that are used in welding operations.

In the context of industrial applications, a previous planning is justified, the so-called *off-line programming*, aiming at establishing a precise control for the movement. This planning includes the analysis of the kinematics and dynamics behavior of the system. The reduction of costs and increase of productivity are some of the most important objectives in industrial automation. Therefore, to make possible the use of robotic systems, it is important that one considers the path planning optimization for a specific task.

The improvement of industrial productivity can be achieved by reducing the weight of the robots and/or increasing their speed of operation. The first choice may lead to power consumption reduction while the second results in a faster work cycle. To successfully achieve these purposes it is very desirable to build flexible robotic manipulators. In some situations it is even necessary to consider the flexibility effects due to the joints and gear components of the manipulators for obtaining an accurate and reliable control.

Compared to conventional heavy robots, flexible link manipulators have the potential advantage of lower cost, larger work volume, higher operational speed, greater payload-to-manipulator-weight ratio, smaller actuators and lower energy consumption.

The study of the control of flexible manipulators started in the field of space robots research. Aiming at space applications, the manipulator should be as light as possible in order to reduce the launching costs (Book, 1984). Uchiyama *et al.* (1990), Alberts *et al.* (1992), Dubowsky (1994), to mention only a few, have also studied flexible manipulators for space applications. Shi *et al.* (1998) discussed some key issues in the dynamic control of lightweight robots for several applications.

As a consequence of the interest in using flexible structures in robotics, several papers regarding the design of controllers for the manipulation task of flexible manipulators are found in the literature (Latornell *et al.*, 1998), (Choi and Krishnamurthy, 1994) and (Chang and Chen, 1998).

In Tsujita *et al.* (2004), the trajectory and force controller of a flexible manipulator is proposed. From the point of view of structural dynamics, the trajectory control for a flexible manipulator is dedicated to the control of the global elastic deformation of the system, and the force control is dedicated to the control of the local deformation at the tip of the end-effector. Thus, preferably trajectory and force controls are separated in the control strategy.

Static and dynamic hybrid position/force control algorithms have been developed for flexible-macro/rigid-micro manipulator systems (Yoshikawa *et al.*, 1996). The robust cooperative control scheme of two flexible manipulators in the horizontal workspace is presented in Matsuno and Hatayama (1999). A passive controller has been developed for the payload manipulation with two planar flexible arms (Damaren 2000).

In Miyabe *et al.* (2004), the automated object capture with a two-arm flexible manipulator is addressed, which is a basic technology for a number of services in space. This object capturing strategy includes symmetric cooperative control, visual servoing, the resolution of the inverse kinematics problem, and the optimization of the configuration of a two-arm redundant flexible manipulator.

The effective use of flexible robotic manipulators in industrial environment is still a challenge for modern engineering. Usually there are several possible trajectories to perform a given task. A question that arises when programming robots is which is the best trajectory. There is no definitive solution, since the answer to this question depends on the selected performance index. Focused on industrial applications, the optimal path planning of a flexible manipulator is addressed in the present chapter. The manipulator is requested to perform a task in a vertical plane. Under this condition the gravitational effects are taken into account. Energy consumption is minimized when the movement is conducted through a suitable path. Energy is calculated by means of the evaluation of the joint torque along the path. End-effector accuracy is improved by reducing the vibration effect and increasing manipulability. The determination of the position takes into account the influence of structural flexibility. Weighting parameters are used to set the importance of each objective. The optimization scenario is composed by an optimal control formulation, solved by means of a nonlinear programming algorithm. The improvement obtained through a global optimization procedure is discussed. Numerical results demonstrate the viability of the proposed methodology.

A control formulation to determine the optimal torque profile is proposed. The optimal manipulability is also taken into account. The effect of using end-effector positioning error as performance index is discussed. As a result, the contribution of the present work is the proposition of a methodology to evaluate the influence of different performance indexes in a multi-criteria optimization environment.

The paper is organized as follows. In Section 2, model of deflection, torque and manipulability are presented. Section 3 recalls the general optimal control formulation and the performance indexes are defined. Geometrical insight about the design variables is given. Multi-criteria programming aspects such as Pareto-optimality and objective weighting are presented in section 4. The global optimization strategy is outlined in section

5 while section 6 shows numerical results. The conclusions and perspectives for future work are given in section 7.

2. Manipulator model

2.1 Deflection

Different schemes for modeling of the manipulators have been studied by a number of researchers. The mathematical model of the manipulator is generally derived from energy principles and, for a simple rigid manipulator, the rigid arm stores kinetic energy due to its moving inertia, and stores potential energy due its position in the gravitational field.

A flexible link also stores deformation energy by virtue of its deflection, joint and drive flexibility. Joints have concentrated compliance that may often be modeled as a pure spring storing only strain energy. Drive components such as shafts and belts may appear distributed. They store kinetic energy due to their low inertia, and a lumped parameter spring model often succeeds well to consider such an effect.

The most important modeling techniques for single flexible link manipulators can be grouped under the following categories: assumed modes method, finite element method and lumped parameters technique.

In the assumed modes approach, the link flexibility is usually represented by a truncated finite modal series, in terms of spatial mode eigenfunctions and time-varying mode amplitudes. Although this method has been widely used, there are several ways to choose link boundary conditions and mode eigenfunctions. Some contributions in this field were presented by Cannon and Schmitz (1984), Sakawa et al. (1985), Bayo (1986), Tomei and Tornambe (1988), among others. Nagaraj et al. (2001), Martins et al. (2002) and Tso et al. (2003) studied single-link flexible manipulators by using Lagrange's equation and the assumed modes method.

Regarding the finite element formulation, Nagarajan and Turcic (1990) derived elemental and system equations for systems with both elastic and rigid links. Bricout et al. (1990) studied elastic manipulators. Moulin and Bayo (1991) also used finite element discretization to study the end-point trajectory tracking for flexible arms and showed that a non-causal solution for the actuating torque enables tracking of an arbitrary tip displacement with any desired accuracy.

By using a lumped parameter model, Zhu et al. (1999) simulated the tip position tracking of a single-link flexible manipulator. Khalil and Gautier (2000) used a lumped elasticity model for flexible mechanical systems. Megahed and Hamza (2004) used a variation of the finite segment multi-body dynamics approach to model and simulate planar flexible link manipulator with rigid tip connections to revolute joints.

Santos et al. (2007) proposed the computation of flexibility by means of a spring-mass-damper system. According to this analogy, the first spring and damper constants are related to the joint behavior, and the following sets of spring and damper represent link flexibility. The variables and the parameters of the model are interpreted as angular quantities.

In this work the description of the deflection related to a rigid link is proposed. It is achieved by means of an Euler-Bernoulli beam formulation and covers the case for small deflections of a beam subject to lateral loads.

The bending moment M, shear forces Q and deflections w for a cantilever beam subjected to a point load P at the free end are given by

$$M(x) = P(x - L) \tag{1}$$

$$Q(x) = P \tag{2}$$

$$w(x) = \frac{Px^2(3L - x)}{6EI} \tag{3}$$

where L is the length of the link, E is the Young's modulus of the material and I is the moment of inertia. The variable x is the distance between the base of the link and the point where the force is applied. In this work, the error of positioning is measured at the end of each link, which yields

$$w(L) = \frac{PL^3}{3EI}. \tag{4}$$

The linear displacement at the end of each link is then converted into angular displacement through the expression

$$\Delta\theta = \arcsin\left(\frac{w}{L}\right). \tag{5}$$

Considering that the kinematics position of each link is given by a rigid body transformation $T_{i-1}^{i}(\theta)$ the positioning error at the end-effector can be estimated through

$$\delta = \sum_{j=1}^{2} T_j(\theta) - T_j(\theta + \Delta\theta) \tag{6}$$

If there is no load at the end of each link, Eq. (4), it follows that $P=0$, $\Delta\theta = 0$ and $\delta = 0$, i.e., the link behaves as a rigid body.

2.2 Robot dynamics

Dynamics encompasses relations between kinematics and statics so that the motion of the system is taken into account. To move a robot through a given trajectory, motors provide force or torque to the robotic joints. There are several techniques to model the dynamics of industrial robots. The knowledge about the dynamic behavior of the system is important both to the computational simulation of the movement and to the design of the controller itself (Fu *et al.* 1987).

Characteristics of a two-link planar manipulator follow. The Denavit-Hartenberg parameters are presented in Table 1.

Link	a (m)	α (rad)	d (m)	θ (rad)
1	a_1	0	0	θ_1^*
2	a_2	0	0	θ_2^*

Table 1. Denavit-Hartenberg parameters, (*) joint variable.

The Lagrangian $L = K - P$ is defined by the difference between the kinetic energy K and the potential energy P of the system. The dynamics of the system can be described by the Lagrange's equations as given by:

$$\tau_j = \frac{d}{dt}\left(\frac{\partial L}{\partial \dot{\theta}_j}\right) - \frac{\partial L}{\partial \theta_j} \tag{7}$$

where θ_j are the generalized coordinates (angle of rotational joints in the present case); $\dot{\theta}_j$ are the generalized velocities (angular velocity in the present case) and τ_j are the generalized forces (torques in the present case). By developing the Euler-Lagrange formalism, the equations that represent the system dynamics are explicitly obtained (Fu *et al.* 1987), and are expressed through

$$Q(q)\ddot{q} + C(q,\dot{q}) + G(q) = \tau \tag{8}$$

where $q = (\theta_1, \theta_2, \ldots, \theta_n)^T$, $\dot{q} = (\dot{\theta}_1, \dot{\theta}_2, \ldots, \dot{\theta}_n)^T$ and $\ddot{q} = (\ddot{\theta}_1, \ddot{\theta}_2, \ldots, \ddot{\theta}_n)^T$ are joint vectors of dimension $n \times 1$ representing position, velocity and acceleration, respectively. $Q(q)$ is a symmetric $n \times n$ matrix representing inertia, $C(q,\dot{q})$ is a $n \times 1$ vector representing the Coriolis effect, $G(q)$ is a $n \times 1$ vector representing the effects of gravitational acceleration, and $F = (F_1, F_2, \ldots, F_n)^T$ is a $n \times 1$ vector of generalized forces applied at the joints.

The Q, C and G matrices that represent the dynamics of a two-link planar manipulator are described by the Eqs. (9) to (16)

$$Q_{11} = \left(\frac{1}{3}\right)m_1\,a_1^2 + \left(\frac{1}{3}\right)m_2\,a_2^2 + m_2\,a_1^2 + m_2\,a_1\,a_2\,cos(\theta_2) \tag{9}$$

$$Q_{12} = \left(\frac{1}{3}\right)m_2\,a_2^2 + 0.5\,m_2\,a_1\,a_2\,cos(\theta_2) \tag{10}$$

$$Q_{21} = \left(\frac{1}{3}\right)m_2\,a_2^2 + 0.5\,m_2\,a_1\,a_2\,cos(\theta_2) \tag{11}$$

$$Q_{22} = \left(\frac{1}{3}\right)m_2\,a_2^2 \tag{12}$$

$$C_{11} = -m_2\,a_1\,a_2\,sin(\theta_2)\dot{\theta}_1\,\dot{\theta}_2 - 0.5\,m_2\,a_1\,a_2\,sin(\theta_2)\,\dot{\theta}_2^{\,2} \tag{13}$$

$$C_{21} = 0.5\,m_2\,a_1\,a_2\,sin(\theta_2)\,\dot{\theta}_2^{\,2} \tag{14}$$

$$G_{11} = -0.5\,m_1\,g\,a_1\,cos(\theta_1) - 0.5\,m_2\,g\,a_2\,cos(\theta_1 + \theta_2) - m_2\,g\,a_1\,cos(\theta_1) \tag{15}$$

$$G_{21} = -0.5\,m_2\,g\,a_2\,cos(\theta_1 + \theta_2) \tag{16}$$

where $g = 9.81 m/s^2$ is the gravitational constant.

The energy required to move the robot is an important design issue because in real applications energy supply is limited and any energy reduction leads to smaller operational costs. This point could lead to eco-robots design, in which energy supply is one of the key aspects. Due to the close relationship that exists between energy and force, the minimal energy can be estimated from the generalized force $\tau_j(t)$ that is associated to each joint j at time instant t.

Furthermore, it was observed that the resulting trajectory corresponds to robot configurations that are associated with the minimum mechanical energy and small variations of the torque along the path. This means that an explicit constraint over the maximum torque is not required since it will be implicitly achieved by the present formulation.

2.3 Manipulability

As an approach for evaluating quantitatively the ability of manipulators from the kinematics viewpoint the concepts of manipulability ellipsoid and manipulability measure are used.

The set of all end-effector velocities v which are realizable by joint velocities such that the Euclidean norm satisfies $|\dot{\theta}| \leq 1$ are taken into account. This set is an ellipsoid in the m-dimensional Euclidean space. In the direction of the major axis of the ellipsoid, the end-effector can move at high speed. On the other hand, in the direction of the ellipse minor axis the end-effector can move only at low speed. Also, the larger the ellipsoid the faster the end-effector can move. Since this ellipsoid represents an ability of manipulation it is called as the manipulability ellipsoid.

The principal axes of the manipulability ellipsoid can be found by making use of the singular-value decomposition of the Jacobian matrix $J(\theta)$. The singular values of J, i.e., $\sigma_1, \sigma_2, \dots, \sigma_m$, are the m larger values taken from the n roots $\left(\sqrt{\lambda_i}, i = 1, \dots, n\right)$ of the eigenvalues. Then λ_i are the eigenvalues of the matrix $J(\theta)^T J(\theta)$.

One of the representative measures for the ability of manipulation derived from the manipulability ellipsoid is the volume of the ellipsoid. This is given by $c_m w$, where

$$w = \sigma_1 . \sigma_2 \dots \sigma_m \tag{17}$$

$$c_m = \begin{cases} (2\pi)^{\frac{m}{2}}/[2.4.6 \dots (m-2).m] & \text{if } m \text{ is even} \\ 2(2\pi)^{\frac{(m-1)}{2}}/[1.3.5 \dots (m-2).m] & \text{if } m \text{ is odd.} \end{cases} \tag{18}$$

Since the coefficient c_m is a constant value when the dimension m is fixed, the volume is proportional to w. Hence, w can be seen as a representative measure, which is called the manipulability measure, associated to the manipulator joint angle θ.

Due to the direct relation between singular configuration and manipulability (through the Jacobian), the larger the manipulability measure the larger the singular configuration avoidance.

The distance from singular points can be obtained from the solution of an optimization problem in which the following objective function is minimized:

$$f(\theta) = \frac{1}{w} \tag{19}$$

Consequently, the minimization of Equation (19) means to increase the manipulability measure.

3. Optimal control formulation

3.1 Design variables

The design variables are the key parameters that are updated during the workflow, aiming at increasing (or decreasing) the performance index. In this study, the design variables are reference nodes, presented by circles in Figure 1. The abscissa presents the time instants $(t_i = 0, t_f = 1)$ while the ordinate is the joint angle. The value of intermediate joint angles between the references nodes are evaluated by means of a cubic spline interpolation.

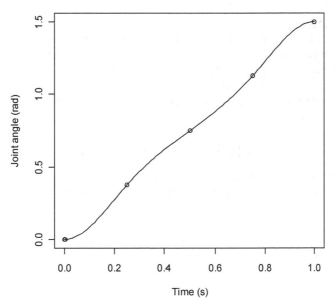

Fig. 1. Reference nodes (o) and joint angle (—).

This approach ensures a smooth transition on the joint angles, velocities and accelerations, which are positive aspects from the mechanical perspective. A smooth movement preserves the mechanism from fatigue effects, for example.

Optimal programming problems for continuous systems are included in the field of the calculus of variations. A continuous-step dynamic system is described by an n-dimensional state vector $\mathbf{x}(t)$ at time t. The choice of an m-dimensional control vector $\mathbf{u}(t)$ determines the time rate of change of the state vector through the dynamics given by the equation below:

$$\dot{\mathbf{x}}(t) = f(\mathbf{x}(t), \mathbf{u}(t), t) \tag{20}$$

A general optimization problem for such a system is to find the time history of the control vector $\mathbf{u}(t)$ for $t_i \leq t \leq t_f$ to minimize a performance index given by

$$J = \varphi[\mathbf{x}(t_f)] + \int_{t_0}^{t_f} L[\mathbf{x}(t), \mathbf{u}(t), t] dt \tag{21}$$

subject to Equation (20) with t_0, t_f, and $\mathbf{x}(t_i)$ specified.

In the present context, the interest is focused on the joint angles and the reference nodes. Performance indexes such as positioning error, manipulability and mechanical power are derived from this information. The variable $x_{i,j} = \theta_j(t_i)$ is an element of the state vector $\mathbf{x}$, which represents the joint angle at each time t_i, related to the joint j. The control vector $u_{k,j}$

represents the position of the reference node k with respect to the joint j. In the present study, the dimension of the state vector is $n = 202$ (the total traveling time is divided into $i=1,...,101$ steps for each joint) and the control vector has dimension $m = 1,...,10$ (five reference nodes for each joint).

3.2 Performance indexes

Initially, the specification of a feasible project is modeled as an optimization problem. To achieve this purpose, let $p_t=[x_{target}, y_{target}]^T$ be the Cartesian target position where the end-effector may intersect when performing a given task. Then, a feasible project is the one whose torque profile is able to conduct each end-effector to a given cartesian position in which the prescribed task will be performed. This profile is obtained by the optimum value of the objective function

$$\varphi[\mathbf{x}(t_f)] = \left[e_1\left(\mathbf{x}(t_f)\right) - p_t\right]^2 \tag{22}$$

where $e_1(\mathbf{x}(t))$ is the Cartesian position of the robot end-effector, respectively. By using Eq. (22) as the only performance index of the general formulation, Eq. (21), the objective function is

$$J = \left[e_1\left(\mathbf{x}(t_f)\right) - p_t\right]^2 \tag{23}$$

The torque required by the actuators to achieve the final position is approximated by a quadratic expression

$$L_1(\mathbf{x}(t), \mathbf{u}(t), t) = \sum_{j=1}^{2} \tau_{i,j}^2 \tag{24}$$

The manipulability index is evaluated by the expression

$$L_2(\mathbf{x}(t), \mathbf{u}(t), t) = \sum_{j=1}^{2} \left(\frac{1}{w_{i,j}^2}\right) \tag{25}$$

The end-effector positioning error along the path is evaluated by using the equation

$$L_3(\mathbf{x}(t), \mathbf{u}(t), t) = \sum_{j=1}^{2} \delta_{i,j}^2 \tag{26}$$

A number of methods are found in the literature to deal with optimal control (OC) problems (Bryson, 1999), (Bertsekas, 1995). In the present contribution, the results are computed through a classical nonlinear programming (NLP) procedure (Betts, 2001). In this case, there is no need of extra parameter computations and the derivatives are numerically evaluated. This choice characterizes a strong point of the proposed methodology since it provides an efficient formulation to solve the optimal control problem addressed.

Since the NLP procedure requires a finite number of points as design variables, the continuity of the OC variables along the time interval is obtained by interpolating the discrete set along the time. Those objectives can be evaluated individually or combined according to the importance of each objective.

4. Multi-criteria programming problem

In multiple-criteria optimization problems one deals with a design variable vector $\mathbf{u}$, which satisfies all constraints and makes as small as possible the scalar performance index that is calculated by taking into account the m components of an objective function vector $J(\mathbf{u})$. This goal can be achieved by the vector optimization problem

$$\min_{\mathbf{u}\in\Omega}\{J(\mathbf{u})|h(\mathbf{u}) = 0, g(\mathbf{u}) \leq 0\} \qquad (27)$$

where $\Omega \subset R^n$ is the domain of the objective function (the design space).

An important feature of such multiple-criteria optimization problems is that the optimizer has to deal with objective conflicts (Eschenauer *et al.*, 1990). Other authors discuss the so-called compromise programming, since there is no unique solution to the problem (Vanderplaats, 1999). In this context, the optimality concept used in this work is presented below.

4.1 Pareto-optimality

A vector $\mathbf{u}^* \in \Omega$ is Pareto optimal for the problem (27) if and only if there are no vector $\mathbf{u} \in \Omega$ with the characteristics:

i. $J_k(\mathbf{u}) \leq J_k(\mathbf{u}^*)$ for all $k \in \{1,...,m\}$.

ii. $J_k(\mathbf{u}) < J_k(\mathbf{u}^*)$ for at least one $k \in \{1,...,m\}$.

For all non-Pareto-optimal vectors, the value of at least one objective function J_k can be reduced without increasing the functional values of the other vector components.

Solutions to multi-criteria optimization problems can be found in different ways by defining the so-called substitute problems. Substitute problems represent different form of obtaining the corresponding scalar objective function (Eschenauer *et al.*, 1990).

4.2 Method of objective weighting

Weighting Objectives is one of the most usual (and simple) substitute models for multi-objective optimization problems (Oliveira and Saramago, 2010). It permits a preference formulation that is independent from the individual minimum for positive weights. The preference function or utility function is here determined by the linear combination of the criteria $J_1, ..., J_m$ together with the corresponding weighting factors $\alpha_1, ..., \alpha_m$ so that

$$p[J(\mathbf{u})] = \sum_{k=1}^{m} \alpha_k J_k(\mathbf{u}), \mathbf{u} \in \Omega. \qquad (28)$$

It is usually assumed that $0 \leq \alpha_k \leq 1$ and $\sum \alpha_k = 1$. It is possible to generate Pareto-optima for the original problem (Equation 27) by varying the weights α_k in the preference function.

4.3 The proposed formulation

Multiple objectives usually have different magnitude values. Numerical comparison among them have no real meaning since they have distinct measurement units and one objective may have a much higher value than another, since they represent different physical quantities. In this context, an initial scaling procedure is justified. In the case of two objective functions considered simultaneously, the following steps are adopted:

1.   Select initial parameters $\mathbf{x}_0(t), \mathbf{u}_0(t), t \in [t_0, t_f]$ and $0 \leq \alpha_1 \leq 1$.

2.   Set $\alpha_2 = 1 - \alpha_1$.

3. Set $\bar{J}_k = J_k(\mathbf{x}_0, \mathbf{u}_0, t)$.
4. Update $\alpha_k = \frac{\alpha_k}{\bar{J}_k}, k = 1,2$.
5. Set $p[J(\mathbf{u})] = \alpha_1 J_1 + \alpha_2 J_2$.

It follows that $J = 1$ at the beginning of the optimization process. At the end, $0 < J < 1$. By using this scaling procedure, the final objective function value provides a non-dimensional index that describes the percentage of improvement. As an example, the final value $J=0.3$ means that the overall objective was reduced to 30% of this initial value.

5. Global optimization

Along the investigation of the optimization problem, there are two kinds of solution points (Luenberger, 1984): *local minimum points*, and *global minimum points*. A point $\mathbf{u}^* \in \Omega$ is said to be a *global minimum point* of f over Ω if $f(\mathbf{u}) \geq f(\mathbf{u}^*)$ for all $\mathbf{u} \in \Omega$. If $f(\mathbf{u}) > f(\mathbf{u}^*)$ for all $\mathbf{u} \in \Omega$, $\mathbf{u} \neq \mathbf{u}^*$, then $\mathbf{u}^*$ is said to be a *strict global minimum point* of f over Ω.

There are several search methods devoted to find the global minimum of a nonlinear objective function, since it is not an easy task. Well known methods such as genetic algorithms (Vose, 1999), differential evolution algorithm (Price *et al.*, 2005) and simulated annealing (Kirkpatrick, 1983) could be used in this case. The main characteristic of these methods is that the global (or near global) optimum is obtained through a high number of functional evaluations.

As proposed by Santos *et al.* (2005), to use the best feature of local optimization method (low computational cost) and global optimization method (global minimum), it is considered using the so-called tunneling strategy (Levy and Gomez, 1985) (Levy and Montalvo, 1985), a methodology designed to find the global minimum of a function. It is composed of a sequence of cycles, each cycle consisting of two phases: a minimization phase having the purpose of lowering the current function value, and a tunneling phase that is devoted to find a new initial point (other than the last minimum found) for the next minimization phase. This algorithm was first introduced in Levy and Gomez (1985), and the name derives from its graphic interpretation.

The first phase of the tunneling algorithm (minimization phase) is focused on finding a local minimum $\mathbf{u}^*$ of Equation (21), while the second phase (tunneling phase) generates a new initial point $\mathbf{u}^0 \neq \mathbf{u}^*$ where $f(\mathbf{u}^0) \leq f(\mathbf{u}^*)$. In summary, the computation evolves through the following phases:

a. Minimization phase: Given an initial point $\mathbf{u}^0$, the optimization procedure computes a local minimum $\mathbf{u}^*$ of $f(\mathbf{u})$. At the end, it is considered that a local minimum is found.
b. Tunneling phase: Given $\mathbf{u}^*$ found above, since it is a local minimum, there exists at least one $\mathbf{u}^0 \in \Omega$, so that $f(\mathbf{u}^0) \leq f(\mathbf{u}^*)$, $\mathbf{u}^0 \neq \mathbf{u}^*$.

In other words, there exists $\mathbf{u}^0 \in Z = \{\mathbf{u} \in \Omega - \{\mathbf{u}^*\} \mid f(\mathbf{u}) \leq f(\mathbf{u}^*)\}$. To move from $\mathbf{u}^*$ to $\mathbf{u}^0$ along the tunneling phase a new initial point $\mathbf{u} = \mathbf{u}^* + \delta$, $\mathbf{u} \in \Omega$ is defined and used in the auxiliary function

$$F(\mathbf{u}) = \frac{f(\mathbf{u}) - f(\mathbf{u}^*)}{[(\mathbf{u} - \mathbf{u}^*)^T (\mathbf{u} - \mathbf{u}^*)]^\gamma} \tag{29}$$

that has a pole in $\mathbf{u}^*$ for a sufficient large value of γ. By computing both phases iteratively, the sequence of local minima leads to the global minimum. Different values for γ are suggested (Levy and Gomez, 1985) (Levy and Montalvo, 1985) for being used in Equation (29) to avoid undesirable points and prevent the search algorithm to fail.

6. Numerical results

Computational evaluation was performed aiming at evaluating the effectiveness of the proposed methodology, as outlined in Figure 2.

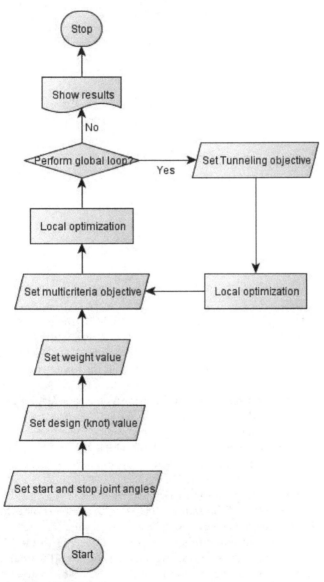

Fig. 2. Computational workflow.

An important point when using multi-objective formulation is the choice of performance indexes. A discussion about the influence of torque, manipulability and end-effector positioning error is presented in the following.

Figure 3 presents the Pareto frontier for the case in which torque and manipulability are considered in the same objective function.

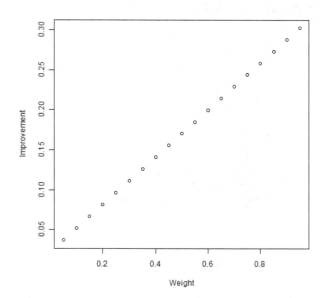

Fig. 3. Minimum value of the performance index.

The abscissa contains several values of the weight factor $\alpha \in (0,1)$. If the factor is zero, only manipulability is taken into account. If the factor is one, only the torque is considered. The factor 0.5 sets the same importance to both objectives. The ordinate is the percentage related to the initial value of the objective function. At the beginning, the objective function value is one. The value 0.10 in the improvement scale means that the final value is 10% of the initial value.

Results presented in the figure confirm that the improvement of manipulability is easier to be achieved by the numerical procedure than the reduction of torque. Those are the final results of the global optimization.

Another important point is the contribution of the global optimization strategy. Figure 4 shows the corresponding improvement achieved.

The abscissa contains several values of the weight factor $\alpha \in (0,1)$. The ordinate represents a range of values where 1.0 means no improvement, that is, no further improvement was obtained with respect to the local minimum. The value 0.75 means that the new (possibly global) minimum has a value corresponding to 75% of the local minimum.

A general view about the contribution of the tunneling strategy to the local minimum is presented in Figure 5. The reduction varies between 0.7 and 1.0, that is, there was a reduction to 70% of the value of the local minimum in the best scenario and no reduction (the corresponding value is 1.0) in the worst case.

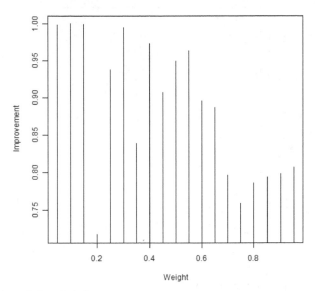

Fig. 4. Contribution of the global strategy.

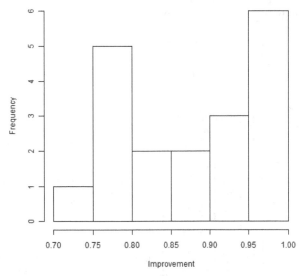

Fig. 5. Range of improvement obtained by the global strategy.

Performance indexes are affected differently by the presence of other objectives and the corresponding priority. The influence of the weight factor on individual objectives is discussed in the following.

The sum of torque is a performance index to be minimized. The higher the importance of the torque in the objective formulation better the improvement achieved by the optimization process. This effect is graphically presented by Figure 6.

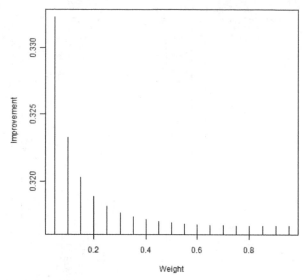

Fig. 6. Influence of weight on the torque.

Manipulability is a performance index to be maximized. It follows that larger the performance index value, better the performance. The influence of the weight coefficient over the manipulability is presented in Figure 7.

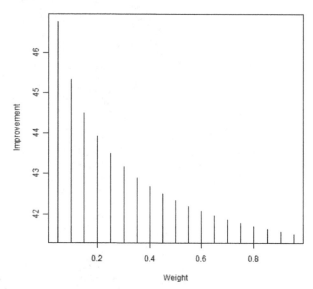

Fig. 7. Influence of weight coefficients on the manipulability.

According to Figure 7, the lower the value of the weight coefficient, better the manipulability. Note that manipulability and torque are conflicting objectives that are addressed by the present formulation.

Results comparing the effects of the end-effector positioning error and torque are presented next.The Pareto frontier for torque and end-effector displacement is shown in Figure 8.

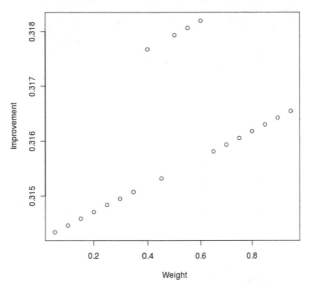

Fig. 8. Minimum value of the performance index.

Despite the small difference in the results, for all cases studied the optimum is between 31% and 32% of the initial value of the objective function. The contribution of the global optimization strategy is also similar for all cases and the global minimum was found between 80.4% and 81.4% of the value of the local minimum.

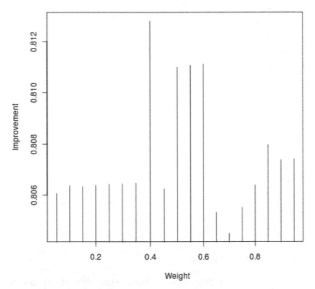

Fig. 9. Contribution of the global strategy.

The average contribution of the tunneling process to the global minimum is shown in Figure 10.

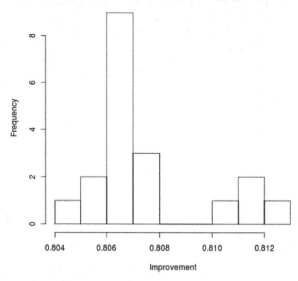

Fig. 10. Range of improvement obtained by the global strategy.

The optimal value of torque index is between 31.6% and 31.9% of the initial value, as presented in Figure 11.

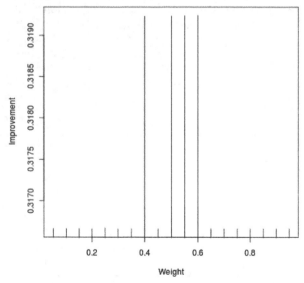

Fig. 11. Influence of weight on the torque.

The end-effector positioning error index was increased about 36 times, as presented in Figure 12.

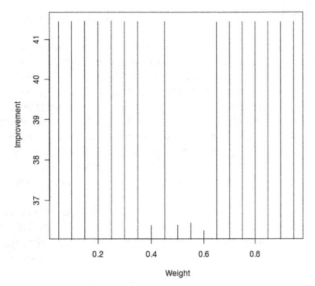

Fig. 12. Influence of weight on the end-effector positioning error.

Results presented in Figures 6 and 7 appear frequently when conflicting objectives are taken into account in the optimization procedure. On the other hand, correlated indexes usually have small deviation in the values obtained, as presented in Figures 11 and 12.

This analysis suggest that torque and manipulability are better choices to compose a multi-criteria analysis as compared with results given by torque and end-effector positioning error. This is justified by the fact that end-effector disturbance is affected by the torque profile. As a result, if the torque is reduced, then the effect of flexibility at the end-effector is also reduced.

7. Conclusion

This work was dedicated to multi-criteria optimization problems applied to flexible robot manipulators. Initially, the model of deflection, torque and manipulability were presented for the system analyzed. Next, optimal control formalism and multi-criteria strategy were outlined. It was concluded that the effectiveness of the numerical procedure depends on the choice of the objective functions and weighting factors.

The first numerical evaluation considered torque and manipulability as performance indexes. The effect of changing the weighting coefficients was presented in Figure 3. This is a typical trend of concurrent objectives, i.e., when one performance index is improved, the other degenerates. In this context it is important to identify the contribution of each index, as shown in Figures 6 and 7.

The effect of the global optimization procedure was also discussed. This point was shown to be effective to obtain the global minimum.

The second numerical evaluation considered the end-effector error positioning and the torque as performance indexes. It was shown that there are no significant changes in the performance index while the weighting factor values are changed. It is explained by the correlation between the objectives, i.e., both indexes are directly affected by the joint torque.

As a result, no further improvement is achieved by changing the relative importance of the objectives.

The numerical experiments conducted during the present research work lead to the following conclusions. The interdependence of the performance indexes is an important aspect that impacts the performance of the numerical method. The evaluation of different objectives that evaluates correlated information increases the computational cost but have no contribution to the optimal design. The use of several weighting factors is recommended. The nonlinear nature of the objective indexes requires a heavy exploration of the design space, aiming at finding a significant improvement for a particular combination of weighing factors. Finally, the use of a global optimization methodology is the most important contribution of this work that will distinguish the proposed methodology as a fast and accurate solver. Few iterations of the tunneling process provided an effective evolution to the global minimum.

From the author's perspective, the effective combination of different techniques is the key to obtain a high performance engineering result. Since this study deals with off-line planning, even a small improvement in the path performed by the robot may lead to expressive economic benefits. This is justified by the fact that the same movement of the robot is repeated several times during the working cycle.

The next step of this research is the evaluation of the effect of uncertainty in the design, through a stochastic approach.

As the proposed methodology is efficient to obtain an improved manipulator trajectory while dealing with the flexibility effect, joint torque and manipulability, the authors believe that the present contribution is a useful tool for robotic path planning design.

8. References

Alberts, T. E., Xia, H. and Chen, Y., 1992, "Dynamic analysis to evaluate viscoelastic passive damping augmentation for the space shuttle remote manipulator system", ASME Journal of Dynamic Systems, Measurement, and Control, Vol. 114, pp. 468–474.

Bayo, E., 1986, "Timoshenko versus Bernoulli beam theories for the control of flexible robots", Proceeding of IASTED International Symposium on Applied Control and Identification, pp. 178–182.

Bertsekas, D. P., 1995, "Dynamic Programming and Optimal Control", Athena Scientific.

Betts, J. T., 2001, "Practical Methods for Optimal Control Using Nonlinear Programming", SIAM 2001, Philadelphia.

Book, W. J., 1984, "Recursive Lagrangian dynamics of flexible manipulator arms", The International Journal of Robotics Research, Vol. 3, No. 3, pp. 87–101.

Bricout, J.N., Debus, J.C. and Micheau, P., 1990, "A finite element model for the dynamics of flexible manipulator", Mechanism and Machine Theory, Vol. 25, No. 1, pp. 119–128.

Bryson, Jr., A. E., 1999, "Dynamic Optimization", Addison Wesley Longman, Inc.

Cannon, R.H. and Schmitz, E., 1984, "Initial experiments on end-point control of a flexible one-link robot", The International Journal of Robotics Research, Vol. 3, No. 3, pp. 62–75.

Chang, Y. C. and Chen, B. S., 1998, "Adaptive tracking control design of constrained robot systems," International Journal of Adaptive Control, Vol. 12, No. 6, pp. 495–526.

Choi, B. O. and Krishnamurthy, K., 1994, "Unconstrained and constrained motion control of a planar 2-link structurally flexible robotic manipulator", Journal of Robotic Systems, Vol. 11, No. 6, pp. 557–571.

Damaren, C. J., 2000, "On the dynamics and control of flexible multibody systems with closed loops", International Journal of Robotics Research, Vol. 19, No. 3, pp. 238–253.

Dubowsky, S., 1994, "Dealing with vibrations in the deployment structures of space robotic systems", in: Fifth International Conference on Adaptive Structures, Sendai International Center, Sendai, Japan, pp. 5–7.

Eschenauer, H. ,Koski, J., Osyczka, A., 1990, "Multicriteria Design Optimization", Berlin, Springer-Verlag.

Fu, K. S., Gonzales, R. C. and Lee, C. S. G., 1987, "Robotics: control, sensing, vision and intelligence", McGraw-Hill.

Khalil, W. and Gautier, M., 2000, "Modeling of mechanical systems with lumped elasticity", Proceedings of the IEEE International Conference on Robotics and Automation, Vol. 4, pp. 3964–3969.

Kirkpatrick, S., Gelatt, C. D. and Vecchi, M. P., 1983, "Optimization by Simulated Annealing", Science, Vol. 220, pp. 671-680.

Latornell, D. J., Cherchas, D. B., and Wong, R., 1998, "Dynamic characteristics of constrained manipulators for contact force control design", International Journal of Robotics Research, Vol. 17, No. 3, pp. 211–231.

Levy, A. V. and Gomez, S., 1985, "The Tunneling Method Ap-plied to Global Optimization. Numerical Optimization," (Ed. P. T. Boggs) R. H. Byrd and R. B. Schnabel, Eds., Society for Industrial and Applied Mathematics, pp. 213-244.

Levy, A. V. and Montalvo, A., 1985, "The Tunneling Algorithm for the Global Minimization of Functions," The SIAM Journal on Scientific and Statistical Computing, Vol. 6, No. 1, 1985, pp. 15-29.

Luenberger, D. G., 1984, "Linear and Nonlinear Programming," 2nd Edition, Addison-Wesley, USA.

Martins, J., Botto, M.A. and Sa da Costa, J., 2002, "Modeling flexible beams for robotic manipulators", Multibody System Dynamics, Vol. 7, No. 1, pp. 79–100.

Matsuno, F. and Hatayama, M., 1999, "Robust cooperative control of two-link flexible manipulators on the basis of quasi-static equations", International Journal of Robotics Research, Vol. 18, No. 4, pp. 414–428.

Megahed, S.M. and Hamza, K.T., 2004, "Modeling and simulation of planar flexible link manipulators with rigid tip connections to revolute joints", Robotica, Vol. 22, pp. 285–300.

Miyabe, T., Konno, A. and Uchiyama, M., 2004, "An Approach Toward an Automated Object Retrieval Operation with a Two-Arm Flexible Manipulator", The International Journal of Robotics Research, Vol. 23, No. 3, March 2004, pp. 275-291.

Moulin, H. and Bayo, E., 1991, "On the accuracy of end-point trajectory tracking for flexible arms by noncausal inverse dynamic solutions", ASME Journal of Dynamic Systems, Measurement, and Control, Vol. 113, pp. 320–324.

Nagaraj, B.P., Nataraju, B.S. and Chandrasekhar, D.N., 2001, "Nondimensional parameters for the dynamics of a single flexible link", in: International Conference on Theoretical, Applied, Computational and Experimental Mechanics (ICTACEM) 2001, Kharagpur, India.

Nagarajan, S. and Turcic, D.A., 1990, "Lagrangian formulation of the equations of motion for the elastic mechanisms with mutual dependence between rigid body and elastic motions, Part 1: Element level equations, and Part II: System equations", ASME Journal of Dynamic Systems, Measurement, and Control, Vol. 112, No. 2, pp. 203–214, and pp. 215–224.

Oliveira, L.S. and Saramago, S.F.P., 2010, "Multiobjective Optimization Techniques Applied to Engineering Problems", Journal of the Brazilian Society of Mechanical Sciences and Engineering, Vol. XXXII, p.94 - 104.

Price, K., Storn, R. and Lampinen, J., 2005, "Differential Evolution - A Practical Approach to Global Optimization", Springer.

Sakawa, Y., Matsuno, F. and Fukushima, F., 1985, "Modelling and feedback control of a flexible arm", Journal of Robotic Systems, Vol. 2, No. 4, pp. 453–472.

Santos, R. R., Steffen Jr., V. and Saramago, S.F.P., 2005, "Solving the inverse kinematics problem through performance index optimization".XXVI Iberian Latin American Congress on Computational Methods in Engineering, Guarapari-ES.CILAMCE.

Santos, R. R., Steffen Jr., V. and Saramago, S.F.P., 2007, "Optimal path planning and task adjustment for cooperative flexible manipulators". 19th International Congress of Mechanical Engineering.

Shi, J.X., Albu-Schaffer, A. and Hirzinger, G., 1998, "Key issues in the dynamic control of lightweight robots for space and terrestrial applications", Proceedings of the IEEE International Conference on Robotics and Automation, Vol. 1, pp. 490–497.

Tomei, P. and Tornambe, A., "Approximate modeling of robots having elastic links", IEEE Transactions on Systems, Man and Cybernetics, Vol. 18, No. 5, pp. 831–840.

Tso, S.K., Yang, T.W., Xu, W.L. and Sun, Z.Q., 2003, "Vibration control for a flexible link robot arm with deflection feedback", International Journal of Non-Linear Mechanics, Vol. 38, pp. 51–62.

Tsujita, K., Tsuchiya, K., Urakubo, T., Sugawara, Z., 2004, "Trajectory and Force Control of a Manipulator With Elastic Links", Journal of Vibration and Control, Vol. 10, pp. 1271–1289.

Uchiyama, M., Konno, A., Uchiyama, T. and Kanda, S., 1990, "Development of flexible dual-arm manipulator tested for space robotics", Proceedings of IEEE International Workshop on Intelligent Robots and System, pp. 375–381.

Vanderplaats, G. N, 1999, "Numerical Optimization Techniques for Engineering Design",3rd edition, VR&D Inc.

Vose, M. D., 1999, "The Simple Genetic Algorithm: Foundations and Theory", MIT Press, Cambridge, MA.

Yoshikawa, T., Harada, K., and Matsumoto, A., 1996, "Hybrid position/force control of flexible-macro/rigid-micro manipulator system", IEEE Transactions on Robotics and Automation, Vol. 12, No. 4, pp. 633–640.

Zhu, G., Ge, S.S. and Lee, T.H., 1999, "Simulation studies of tip tracking control of a single-link flexible robot based on a lumped model", Robotica, Vol. 17, pp. 71–78.

8

Singularity Analysis, Constraint Wrenches and Optimal Design of Parallel Manipulators

Nguyen Minh Thanh[1], Le Hoai Quoc[2] and Victor Glazunov[3]
[1]*Department of Automation, Hochiminh City University of Transport,*
[2] *Department of Science and Technology, People's Committee of Hochiminh City,*
[3]*Mechanical Engineering Research Institute, Russian Academy of Sciences,*
[1,2]*Vietnam*
[3]*Russia*

1. Introduction

In recent years, numerous researchers have investigated parallel manipulators and many studies have been done on the kinematics or dynamics analysis. Parallel manipulators has been only mentioned in several books, as in (Merlet, 2006; Ceccarelli, 2004; Kong, & Gosselin, 2007; Glazunov, et al., 1991). Reference (Gosselin, & Angeles, 1990) has established singularity criteria based on Jacobian matrices when describing the various types of singularity. Then, in (Glazunov, et al. 1990) proposed other singularity criteria for consideration of these problems the screw theory based on the approach of the screw calculus, as in (Dimentberg, 1965). Those criteria are determined by the constraints imposed by the kinematic chains, as in (Angeles, 2004; Kraynev, & Glazunov, 1991), taking into account some problems the Plücker coordinates of constraint wrenches can be applied in (Glazunov, 2006; Glazunov, et al. 1999, 2007, 2009; Thanh, et al. 2009, 2010a).

Dynamical decoupling allows increasing the accuracy for the parallel manipulators presented as in (Glazunov, & Kraynev, 2006; Glazunov, & Thanh, 2008). It is necessary to develop optimal structure have combined (Thanh, et al. 2008), as well as algorithms and multi-criteria optimization (Statnikov, 1999; Thanh, et al. 2010b) obtaining the Pareto set. It is very important to taking into account possible singularity configurations, to find out how they influence the characteristics of constraints restricting working space (Bonev, et al. 2003; Huang, 2004; Arakelian, et al. 2007).

The trend towards highly rapid manipulators due to the demand for greater working volume, dexterity, and stiffness has motivated research and development of new types of parallel manipulator (Merlet, 1991). This paper is focused the constraints and criteria existing in known parallel manipulators in form of a parallel manipulator with linear actuators located on the base.

2. Kinematic of parallel manipulator

In this section, let us consider a 6-DOF parallel manipulator with actuators situated on the base. The mechanical architecture of the considered robot is illustrated in Fig. 1.

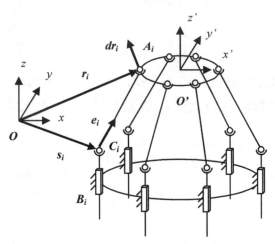

Fig. 1. Parallel manipulator with linear actuators located on the base

The parallel manipulator as seen in Fig. 1 is composed of a mobile platform connected to a fixed base via six kinematic sub-chains (legs) comprising of one prismatic, one universal and one spherical pair (PUS pairs). Parameters of design of the platform and the base form an irregular hexagon positioned in the $(x$-$y)$ plane. A_i, B_i $(i=1,\ldots,6)$ are coordinates of the points of the mobile platform (the output link) and of the base respectively. The points $A_1A_3A_5$ and $A_2A_4A_6$ make form equilateral triangles, the angle ψ_p determines their location and R_p is the radius of the circumscribed circle (Fig. 2, a). Similarly, the angle ψ_b and the radius R_b determines the location of the equilateral triangles $B_1B_3B_5$ and $B_2B_4B_6$ located on the base. Let the distance between the centers of the universal and spherical pairs A_i and C_i of the i-th leg be l_i. In addition, the generalized coordinates, which are equal to the distance between the points B_i and C_i are designated θ_i. The radius-vectors of the points A_i and C_i are $r_i(x_{Ai}, y_{Ai}, z_{Ai})$ and $s_i(x_{Ci}, y_{Ci}, z_{Ci})$ respectively $(i=1,\ldots,6)$. We could note that the coordinates of the points B_i and C_i are $x_{Ci}=x_{Bi}$, $y_{Ci}=y_{Bi}$, $z_{Ci}=\theta_i$.

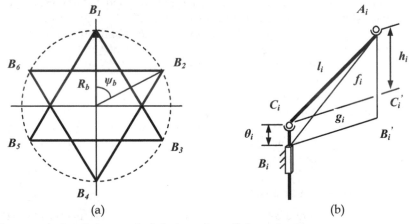

(a)   (b)

Fig. 2. Parametrical and geometrical design of parallel manipulator

With this approach, the linear actuators can be firmly fixed on the base to reduce high acceleration movements because the power is not used to move heavy actuators but lightweight links. However, the obstacle is a smaller working space in comparison with a Stewart platform, due to the movement of the linear actuators. Moreover, forces acting on the actuator have a perpendicular component, whereas forces exerted upon Stewart actuators have a longitudinal component.

Let us consider an inverse kinematic problem of position of parallel manipulators, which has characteristic relation between the numbers of chains. The manipulator with six kinematic chains offers convenience in optimization of working space in terms of decreased rigidity and load-bearing capacity.

Likewise, the generalized coordinates of the *i-th* segment (the length) which are equal to the distance between the points A_i and B_i, can be expressed as:

$$f_i = \sqrt{\left(x_{Ai} - x_{Bi}\right)^2 + \left(y_{Ai} - y_{Bi}\right)^2 + \left(z_{Ai} - z_{Bi}\right)^2} , i = 1,...,6 \tag{1}$$

By geometrical method, the distance between the points C_i and C'_i (Fig. 2, b):

$$g_i = \sqrt{(f_i)^2 - (z_{Ai})^2} , i = 1,...,6 \tag{2}$$

when the length of the link l_i is known, the distance between the points A_i and C'_i (Fig. 2, b):

$$h_i = \sqrt{(l_i)^2 - (g_i)^2} \tag{3}$$

We could obtain the generalized coordinate θ_i (Fig. 2, c) as follows:

$$\theta_i = z_{Ai} - h_i \tag{4}$$

It is the solution of the inverse kinematic problem. The inverse kinematics for parallel manipulator can be formulated to determine the required actuator heights for a given pose of the mobile platform with respect to the base. The pose consists of both position and orientation in the Cartesian system. Actuators are considered to act linearly in the vertical direction, parallel to the z-axis, in order to simplify the mathematics, although that needs not be the case.

3. Multi-criteria optimization

Influence of singularities on parameters of the working space of the parallel manipulator is a significant factor worth investigating. In these singularity configurations, the system is out of control and that greatly affects its functionality. It is necessary to determine the extent of the lack of control to see how that affects the parameters of the working space. These singularity configurations also affect the optimization results.

The constraint wrenches of zero pitch acting to the output link from the legs are located along the unit screws: $E_i = e_i + \chi e^o_i$, $(i=1,...,6)$ where e_i is the unit vector directed along the axis of the line C_iA_i of the corresponding leg, χ is the Clifford factor, $\chi^2=0$ (for a vector, $e_i e^o_i = 0$). E_i consists of the unit vector e_i and its moment $e^o_i = s_i \times e_i$ corresponding with $e^o_{xi} = s_{Cyi} e_{Czi} - s_{Czi} e_{Cyi}$; $e^o_{yi} = s_{Czi} e_{Cxi} - s_{Cxi} e_{Czi}$; $e^o_{zi} = s_{Cxi} e_{Cyi} - s_{Cyi} e_{Cxi}$ and can be expressed by Plücker coordinates $E_i = (x_i, y_i, z_i, x^o_i, y^o_i, z^o_i)$. These coordinates make form the 6×6 matrix (E):

$$(E) = \begin{pmatrix} x_1 & y_1 & z_1 & x_1^o & y_1^o & z_1^o \\ x_2 & y_2 & z_2 & x_2^o & y_2^o & z_2^o \\ \cdot & \cdot & \cdot & \cdot & \cdot & \cdot \\ \cdot & \cdot & \cdot & \cdot & \cdot & \cdot \\ \cdot & \cdot & \cdot & \cdot & \cdot & \cdot \\ x_6 & y_6 & z_6 & x_6^o & y_6^o & z_6^o \end{pmatrix} \qquad (5)$$

Optimization of parameters of the parallel manipulator with linear actuators located on the base is considered. Let us take in account three criteria: working volume, dexterity and stiffness of parallel manipulator. The first criterion N_p is the quantity of the reachable points of the centre of the mobile platform. The second criterion N_c is the average quantity of orientations of the mobile platform in each reachable point. The third criterion D_e is the average module of the determinant $|\det(E)|$ in each configuration. Determinant $|\det(E)|$ constructed from coordinate axes of the drive kinematic couples is used as a third criterion of optimization. Since the value of this criterion is related to one of the important characteristics of the manipulator - its stiffness or load capacity. If determinant are more qualifiers, then the manipulator away from the singularity configuration and the stiffness of the above.

Let us consider optimization of the parameters of the manipulator for different values of the criterion of proximity to singular configurations, as well as the influence of this criterion in the optimization results. We set up four coefficients H1, H2, H3 and H4 expressed four parameters of optimization. The coefficient H1 characterizes the length $l = l_i$ of the links A_iC_i ($i=1,\dots,6$) (in Fig. 2, b). The coefficient H2 characterizes the angle ψ_p (Fig. 2, a) determining the location of the triangles $A_1A_3A_5$ and $A_2A_4A_6$ of the mobile platform. The coefficient H3 characterizes the angle ψ_b determining the location of the triangles $B_1B_3B_5$ and $B_2B_4B_6$ on the base. Moreover, the coefficient H4 characterizes the relation between the radius R_p and the radius R_b of the circumscribe circles of the platform and of the base respectively.

The algorithm of determination of the Pareto-optimal solutions can be presented as follows:

Step 1. Establish the limits of the parameters of optimization.

$H_{1min} \leq H_1 \leq H_{1max}, H_{2min} \leq H_2 \leq H_{2max}, H_{3min} \leq H_3 \leq H_{3max}, H_{4min} \leq H_4 \leq H_{4max}$. The number of steps of scanning in the space of parameters is n_p. The limits of the scanned Cartesian coordinates of the centre of the moving platform and the limits of the scanned orientation angles of this platform in interval are $x_{min} \leq x \leq x_{max}, y_{min} \leq y \leq y_{max}, z_{min} \leq z \leq z_{max}, \alpha_{min} \leq \alpha \leq \alpha_{max}, \beta_{min} \leq \beta \leq \beta_{max}, \gamma_{min} \leq \gamma \leq \gamma_{max}$. As well as the number n_c of steps of scanning in the space of these coordinates and the limitation of changing of the generalized coordinates $\theta_{imin} \leq \theta_i \leq \theta_{i\,max}$ ($i=1,\dots,6$). The limit of the determinant $|\det(E)| \geq \varepsilon$. At this step assume $i=0$, by this the parameters are $H_{10} = H_{1min}, H_{20} = H_{2min}, H_{30} = H_{3min}, H_{40} = H_{4min}, N_p = N_c = D_e = 0$.

Step 2. Determine the values of the criteria for all the values of the parameters.

2.1. Determine the parameters $H_{1i},\dots, H_{40}$, assume $j=0$, by this $x_0 = x_{min}, y_0 = y_{min}, z_0 = z_{min}, \alpha_0 = \alpha_{min}, \beta_0 = \beta_{min}, \gamma_0 = \gamma_{min}$.

2.2. Determine θ_i ($i=1,\dots,6$) and $|\det(E)|$; if all the $\theta_i\,_{min} \leq \theta_i \leq \theta_i\,_{max}$ ($i=1,\dots,6$) and $|\det(E)| \geq \varepsilon$ then $N_c = N_c + 1$, $D_e = D_e + |\det(E)|$; if $x_j \neq x_{j-1}$ or $y_j \neq y_{j-1}$ or $z_j \neq z_{j-1}$ then $N_p + 1$.

2.3. $j = j+1$, if $j \le n_c$ then go back to 2.2.

2.4. Determine the criteria $N_{pi} = N_p$, $D_{ei} = D_e / N_c$, $N_{ci} = N_c / N_p$; assume $N_p = N_c = D_e = 0$.

2.5. $i = i+1$, if $i \le n_p$ then go back to 2.1.

Step 3. Determine the Pareto-optimal solutions (matrix *(Par)*).

3.1. Assume i= 1, (by this $N_{pi} = N_{p1}$, $D_{ei} = D_{e1}$, $N_{ci} = N_{c1}$), k= 0.

3.2. Determine N_{pi}, D_{ei}, N_{ci}; assume *j= 1*, the criteria of optimal solution *K1=1, K2=0*.

3.3. Determine N_{pj}, D_{ej}, N_{cj}; if $N_{pi} > N_{pj}$ or $D_{ei} > D_{ej}$ or $N_{ci} > N_{cj}$ then K2=1; if $N_{pi} = N_{pj}$ and $D_{ei} = D_{ej}$ and $N_{ci} = N_{cj}$ then K2=1.

3.4. If *K2≠1* then *K1=0; K2=0; j=j+1*; if $j \le n_p$ then go back to 3.3.

3.5. If *K1 = 1* then $k = k + 1$, $Par_{1k} = H_{1i}$, ..., $Par_{7k} = N_{ci}$, *i=i+1*; if $i \le n_p$ then go back to 3.2.

Singularity of the manipulator is determined by closeness to zero of determinant of matrix *(E)* of Plücker coordinates of unit wrenches. Let us fix certain value $\varepsilon > 0$ as a criterion of singularity (the manipulator is in singular position if $|\det(E)| \le \varepsilon$). If $\varepsilon = 0$ then the construction of the working space of the manipulator shows that the same results we can get without singularity constraint. Giving various values of the criterion of the singularity, we can get interval of the determinant of matrix *(E)*.

Further, analysis influences of the criterion of singularity ε, $|\det(E)| \le \varepsilon$ on the value of the working volume. With the value of the criterion of singularity is equal to *ε=0.01* there exist *81* available solutions, but only *8* of them are Pareto-optimal. By the condition of the criterion of singularity is equal to *ε=0.01* and the condition *ε=0*, there Pareto-optimal set consists of *6* and *29* solutions correspondingly. Therefore, the value of ε influences on the results of optimization.

Value of the criterion that determines the proximity to singular configurations is equal to zero, we can assume that the constraints associated with the singularity in general, are not imposed in the analysis of each specific configuration. However, the criterion for determining the load capacity occurs. As a result, the number of Pareto-optimal variants varies very much. Here, *29* variants satisfy the conditions of Pareto set.

Limiting possible module of a determinant of matrix *(E)* to singularity configurations changes the Plücker coordinates of the wrenches transmitted on the output link. Methodology for analyzing the singularities on optimization appearing in the parallel manipulator and their impact in the working space is proposed. The practical significance from the fact is the results obtained in this work increase the effectiveness of design automation.

4. Twist inside singularity

The approach based on matrix *(E)* consisting of the Plücker coordinates of the constraint wrenches allows determining the twists of the platform inside singularity (Glazunov, 2006). Let us consider the increases of the Plücker coordinates of the unit screws E_i after an infinitesimal displacement $\$ = (d\varphi, dr) = (dr_x, dr_x, dr_x, d\varphi_x, d\varphi_y, d\varphi_z)^T$ of the platform corresponding to displacement $dr_i = (dx_{Ai}, dy_{Ai}, dz_{Ai})^T$ of the point A_i of the manipulator presented on the Fig. 1.

$$dx_{Ai} = dr_x + d\varphi_y z_{Ai} - d\varphi_z y_{Ai},$$
$$dy_{Ai} = dr_y + d\varphi_z x_{Ai} - d\varphi_x z_{Ai}, \qquad (6)$$
$$dz_{Ai} = dr_z + d\varphi_x y_{Ai} - d\varphi_y x_{Ai}$$

The generalized coordinate after mentioned infinitesimal displacement (dx_i, dy_i, dz_i) is:

$$\theta_i + d\theta_i = z_{Ai} + dz_{Ai} - \sqrt{l_i^2 - \left(x_{Ai} + dx_{Ai} - x_{Ci}\right)^2 - \left(y_{Ai} + dy_{Ai} - y_{Ci}\right)^2} \tag{7}$$

After transformations the increase of the generalized coordinate is:

$$d\theta_i = \frac{\left[\left(x_{Ai} - x_{Ci}\right)dx_{Ai} + \left(y_{Ai} - y_{Ci}\right)dy_{Ai} + \left(z_{Ai} - z_{Ci}\right)dz_{Ai}\right]}{\left(z_{Ai} - z_{Ci}\right)} \tag{8}$$

The unit screw E_i can be rewritten as $E_i + dE_i$ or as $e_i + de_i$ and $e^o_i + de^o_i$.
Using (6), (7), and (8) the coordinates of the de_i and de^o_i can be expressed as:

$$dx_i = \frac{\partial x_i}{\partial \varphi_x}d\varphi_x + \frac{\partial x_i}{\partial \varphi_y}d\varphi_y + \frac{\partial x_i}{\partial \varphi_z}d\varphi_z + \frac{\partial x_i}{\partial r_x}dr_x + \frac{\partial x_i}{\partial r_y}dr_y + \frac{\partial x_i}{\partial r_z}dr_z$$

$$\cdots\cdots\cdots\cdots\cdots\cdots\cdots\cdots\cdots\cdots\cdots\cdots\cdots\cdots\cdots\cdots\cdots\cdots\cdots\cdots \tag{9}$$

$$dz_i^o = \frac{\partial z_i^o}{\partial \varphi_x}d\varphi_x + \frac{\partial z_i^o}{\partial \varphi_y}d\varphi_y + \frac{\partial z_i^o}{\partial \varphi_z}d\varphi_z + \frac{\partial z_i^o}{\partial r_x}dr_x + \frac{\partial z_i^o}{\partial r_y}dr_y + \frac{\partial z_i^o}{\partial r_z}dr_z$$

where

$$\frac{\partial x_i}{\partial \varphi_x} = 0, \quad \frac{\partial x_i}{\partial \varphi_y} = \frac{z_{Ai}}{l_i}, \quad \frac{\partial x_i}{\partial \varphi_z} = -\frac{y_{Ai}}{l_i}, \quad \frac{\partial x_i}{\partial r_x} = \frac{1}{l_i}, \quad \frac{\partial x_i}{\partial r_y} = 0, \quad \frac{\partial x_i}{\partial r_z} = 0,$$

$$\frac{\partial x_i^o}{\partial \varphi_x} = \frac{\left\{\left(y_{Ai} - y_{Ci}\right)\left[\dfrac{\left(y_{Ai} - 2y_{Ci}\right)z_{Ai}}{\left(z_{Ai} - z_{Ci}\right)} - y_{Ai}\right] + z_{Ci}z_{Ai}\right\}}{l_i},$$

$$\frac{\partial x_i^o}{\partial \varphi_y} = \frac{\left[\dfrac{\left(x_{Ai} - x_{Ci}\right)\left(2y_{Ci} - y_{Ai}\right)z_{Ai}}{\left(z_{Ai} - z_{Ci}\right)} - x_{Ai}\left(y_{Ai} - y_{Ci}\right)\right]}{l_i},$$

$$\frac{\partial x_i^o}{\partial \varphi_z} = \frac{\left\{\dfrac{\left[\left(y_{Ai} - y_{Ci}\right)x_{Ai} - \left(x_{Ai} - x_{Ci}\right)y_{Ai}\right]\left(2y_{Ci} - y_{Ai}\right)}{\left(z_{Ai} - z_{Ci}\right)} - x_{Ai}z_{Ci}\right\}}{l_i},$$

$$\frac{\partial x_i^o}{\partial r_x} = \frac{\left(x_{Ai} - x_{Ci}\right)\left(2y_{Ci} - y_{Ai}\right)}{\left(z_{Ai} - z_{Ci}\right)l_i}, \quad \frac{\partial x_i^o}{\partial r_y} = \frac{\dfrac{\left(y_{Ai} - y_{Ci}\right)\left(2y_{Ci} - y_{Ai}\right)}{\left(z_{Ai} - z_{Ci}\right)} - z_{Ci}}{l_i},$$

$$\frac{\partial x_i^o}{\partial r_z} = \frac{\left(y_{Ci} - y_{Ai}\right)}{l_i}, \quad i = 1,...,6$$

Other partial derivatives also can be obtained from Eqs. (6), (7), and (8).
By means of the properties of linear decomposition of determinants d[det(E)] can be obtained as the sum of 36 determinants (Glazunov, 2006). From this, d[det(E)] can be presented as:

$$d[\det(E)] = \frac{\partial[\det(E)]}{\partial\varphi_x d\varphi_x} + \frac{\partial[\det(E)]}{\partial\varphi_y d\varphi_y} + \frac{\partial[\det(E)]}{\partial\varphi_z d\varphi_z} +$$
$$+ \frac{\partial[\det(E)]}{\partial r_x dr_x} + \frac{\partial[\det(E)]}{\partial r_y dr_y} + \frac{\partial[\det(E)]}{\partial r_z dr_z}$$

(10)

Using (10) the criterion of the singularity locus can be presented as d[det(E)]=0. This condition imposes only one constraint. Therefore, there exist five twists of motions of the platform inside singularity.

For example, let us obtain five inside singularity of manipulator. Set up the coordinates of the vectors be r_i are $r_1(-1, 0, 4)$, $r_2(-0.5, 1, 4)$, $r_3(0.5, 1, 4)$, $r_4(1, 0, 4)$, $r_5(0.5, -1, 4)$, $r_6(-0.5, -1, 4)$; s_i are $s_1(-1.5, 0, 0.866)$, $s_2(-1, 1.5, 0.707)$, $s_3(1, 1.5, 0.707)$, $s_4(1.5, 0, 0.866)$, $s_5(1, -1.5, 0.707)$, $s_6(-1, -1.5, 0.707)$. From here, we can see the generalized coordinates as (0.136, 0.305, 0.305, 0.136, 0.305, 0.305). Matrix (E) is determined as:

$$\begin{pmatrix}
0.158 & 0 & 0.988 & 0 & 1.618 & 0 \\
0.148 & -0.148 & 0.978 & 1.572 & 1.083 & -0.074 \\
-0.148 & -0.148 & 0.978 & 1.572 & -1.083 & 0.074 \\
-0.158 & 0 & 0.988 & 0 & -1.618 & 0 \\
-0.148 & 0.148 & 0.978 & -1.572 & -1.083 & -0.074 \\
0.148 & 0.148 & 0.978 & -1.572 & 1.083 & 0.074
\end{pmatrix}$$

The determinant consisting of the Plücker coordinates of the unit screws is det(E) = 0. Their partial derivatives are:

$$\frac{\partial[\det(E)]}{\partial r_x} = -0.02, \quad \frac{\partial[\det(E)]}{\partial r_y} = 0, \quad \frac{\partial[\det(E)]}{\partial r_z} = 0,$$

$$\frac{\partial[\det(E)]}{\partial\varphi_x} = 0, \quad \frac{\partial[\det(E)]}{\partial\varphi_y} = 0.002, \quad \frac{\partial[\det(E)]}{\partial\varphi_z} = 0$$

Using the approach presented above, we find five independent twists inside singularity: $1(1, 0, 0, 0, 0, 0)$, $2(0, 0, 1, 0, 0, 0)$, $3(0, 0, 0, 0, 1, 0)$, $4(0, 0, 0, 0, 0, 1)$, $5(0, 4.433, 0, 1, 0, 0)$. The twist-gradient is calculated to be $*(-0.02, 0, 0, 0, 0.002, 0)$. This twist-gradient is practically important as it offers the highest speed.

5. Dynamical decoupling

In this section, let we consider the reduction of the dynamical coupling of the motors of the parallel manipulator with linear actuators located on the base. The basic idea is to represent the kinetic energy as the quadratic polynomial including only the squares of the generalized velocities (Glazunov, & Kraynev, 2006). The kinetic energy can be expressed by means of the matrix (E).

Let m be the mass and J_x, J_y, J_z be the inertia moments of the platform. Assuming that the mass of the platform is much more than the masses of the legs and using the Eqs. (6)-(8), the kinetic energy T can be expressed as follows (Dimentberg, 1965; Kraynev, & Glazunov, 1991):

$$T = \frac{m}{2}\left[\left(\sum_{i=1}^{6} p_i^0 \dot{\theta}_i G_{ii}\right)^2 + \left(\sum_{i=1}^{6} q_i^0 \dot{\theta}_i G_{ii}\right)^2 + \left(\sum_{i=1}^{6} r_i^0 \dot{\theta}_i G_{ii}\right)^2\right] +$$

$$(11)$$

$$+ \frac{J_x}{2}\left(\sum_{i=1}^{6} p_i \dot{\theta}_i G_{ii}\right)^2 + \frac{J_y}{2}\left(\sum_{i=1}^{6} q_i \dot{\theta}_i G_{ii}\right)^2 + \frac{J_z}{2}\left(\sum_{i=1}^{6} r_i \dot{\theta}_i G_{ii}\right)^2$$

where $\dot{\theta}_i$ are the generalized velocities, p_i, q_i, r_i, p^0_i, q^0_i, r^0_i are the components of the matrix $(E)^{-1}$, G_{ii} are the components of the diagonal matrix (G) $(i=1,\ldots,6)$.

The Lagrange equation of motion for a parallel manipulator can be written as:

$$\frac{d}{dt}\left(\frac{\partial T}{\partial \dot{\theta}_i}\right) - \frac{\partial T}{\partial \theta_i} = Q_i \qquad (12)$$

where Q_i are the generalized forces $(i=1,\ldots,6)$.

The dynamical coupling can be determined using the Eq. (11). The expression of each generalized force comprises all other generalized velocities and accelerations. In order to reduce the dynamical coupling we represent the kinetic energy as follows:

$$T = \frac{m}{2}\left[\begin{array}{c}\sum\limits_{i=1}^{6}\left(p_i^0 \dot{\theta}_i G_{ii}\right)^2 + 2\sum\limits_{i=1,j=1,i\neq j}^{6} p_i^0 p_j^0 \dot{\theta}_i G_{ii} \dot{\theta}_j G_{jj} + \\[2ex] +\ldots+ \sum\limits_{i=1}^{6}\left(r_i^0 \dot{\theta}_i G_{ii}\right)^2 + 2\sum\limits_{i=1,j=1,i\neq j}^{6} r_i^0 r_j^0 \dot{\theta}_i G_{ii} \dot{\theta}_j G_{jj}\end{array}\right] +$$

$$+ \frac{J_x}{2}\left[\sum\limits_{i=1}^{6}\left(p_i \dot{\theta}_i G_{ii}\right)^2 + 2\sum\limits_{i=1,j=1,i\neq j}^{6} p_i p_j \dot{\theta}_i G_{ii} \dot{\theta}_j G_{jj}\right] + \qquad (13)$$

$$+\ldots+ \frac{J_z}{2}\left[\sum\limits_{i=1}^{6}\left(r_i \dot{\theta}_i G_{ii}\right)^2 + 2\sum\limits_{i=1,j=1,i\neq j}^{6} r_i r_j \dot{\theta}_i G_{ii} \dot{\theta}_j G_{jj}\right]$$

According to the Eq. (13) dynamical decoupling can be satisfied if the columns of the following matrix (D) are orthogonal:

$$(D) = \begin{pmatrix} p_1^0 G_{11}\sqrt{m} & . & . & . & . & p_6^0 G_{66}\sqrt{m} \\ q_1^0 G_{11}\sqrt{m} & . & . & . & . & q_6^0 G_{66}\sqrt{m} \\ r_1^0 G_{11}\sqrt{m} & . & . & . & . & r_6^0 G_{66}\sqrt{m} \\ p_1 G_{11}\sqrt{J_x} & . & . & . & . & p_6 G_{66}\sqrt{J_x} \\ q_1 G_{11}\sqrt{J_y} & . & . & . & . & q_6 G_{66}\sqrt{J_y} \\ r_1 G_{11}\sqrt{J_z} & . & . & . & . & r_6 G_{66}\sqrt{J_z} \end{pmatrix}$$

$$(D) = \begin{pmatrix} \sqrt{m} & 0 & 0 & 0 & 0 & 0 \\ 0 & \sqrt{m} & 0 & 0 & 0 & 0 \\ 0 & 0 & \sqrt{m} & 0 & 0 & 0 \\ 0 & 0 & 0 & \sqrt{J_x} & 0 & 0 \\ 0 & 0 & 0 & 0 & \sqrt{J_y} & 0 \\ 0 & 0 & 0 & 0 & 0 & \sqrt{J_z} \end{pmatrix} (E)^{-1}(G)$$

$$(D) = (M)(E)^{-1}(G) \tag{14}$$

Let the axes of the links A_iC_i be parallel to the axes of the links B_iC_i and the matrix (G) is unit matrix. Then in order to satisfy the condition of orthogonal columns of the matrix (D) the rows of the inverse matrix $(D)^{-1} = (E)(M)^{-1}$ are to be orthogonal. From this, the following matrix (E) is proposed:

$$(E) = \begin{pmatrix} 0 & 0 & 1 & 0 & -\sqrt{J_y/m} & 0 \\ 0 & 0 & -1 & 0 & -\sqrt{J_y/m} & 0 \\ 1 & 0 & 0 & 0 & 0 & -\sqrt{J_z/m} \\ -1 & 0 & 0 & 0 & 0 & -\sqrt{J_z/m} \\ 0 & 1 & 0 & -\sqrt{J_x/m} & 0 & 0 \\ 0 & -1 & 0 & -\sqrt{J_x/m} & 0 & 0 \end{pmatrix} \tag{15}$$

The determinant of the matrix (E) (15) can be written as:

$$\det(E) = 8\sqrt{J_x J_y J_z /m^3} \tag{16}$$

The matrix (15) corresponds to the Fig. 3. Here the center of the mass of the platform coincides with the center of the coordinate system xyz and the axes of the links of the legs

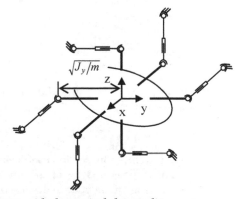

Fig. 3. Parallel manipulator with dynamical decoupling

are parallel to the main central inertia axes of the platform. The proposed approach can be applicable for manipulators characterized by small displacements and high speeds. Moreover, this architecture causes partial kinematic decoupling because if the generalized coordinates corresponding to the opposite legs are equivalent then the moving platform keeps constant orientation.

6. Pressure angles

The parallel manipulators have singularity configurations in which there is an uncontrolled mobility because some of the wrenches acting on the output link are linearly dependent. The local criterion of singular configurations is the singular matrix of the screw coordinates of the wrenches, such as:

$$\det(E) = \varepsilon^* \tag{17}$$

where ε^* is the preassigned minimal determinant value. The pressure angle of the linear dependent sub-chain is equal to $\pi/2$, as a reciprocal twist to five-member group of screws has a perpendicular moment at about any points of the axis. All stalled actuators but one the manipulator has $DOF=1$ and its output link can move along some twist $\Omega = \omega + \chi\omega^0$ $(\chi^2 = 0)$ reciprocal to five-member group of the wrenches corresponding to stalled actuators. We can find this twist from the reciprocity condition:

$$mom(\Omega, R_i) = 0, \ i = 1,...,5 \tag{18}$$

In general the six-member group of the unit wrenches of zero parameter $R_i(r_i, r_i^o)$ $(i=1,...,6)$ is acting on the output link of the such manipulators, determinant composed of the screw coordinates of these wrenches as given in (5).
The velocity of any point A_i $(i=1,...,6)$ of the mobile platform can be found as a twist moment relative to this point:

$$V_{A_i} = \omega^0 + \omega \times r_{A_i}, \ i = 1,...,6 \tag{19}$$

where r_{Ai} is radius-vectors of the points A_i.
The pressure angle α_i for the stalled actuator i-th of the parallel manipulators (Fig.1) can be determined as:

$$\alpha_i = \arccos\left(\frac{V_{A_i} \cdot F_i}{\left|V_{A_i}\right| \cdot \left|F_i\right|}\right), \ i = 1,...,6 \tag{20}$$

where F_i is the force vector on the actuator axis. For normal functions of the manipulator it is necessary that working space be limited by positions:

$$\alpha_i \leq \alpha_{KP}, \ i = 1,...,6 \tag{21}$$

where α_{KP} is maximum pressure angle is defined by friction coefficient.
The manipulator control system must be provided by algorithm testing the nearness to singular configurations based on the analysis of singular matrix (5) or on the pressure angle determination.

7. Manipulator for external conditions

In Fig. 4, (a, b) the six-DOFs parallel mechanisms and their sub-chain has a parallel connection of links and actuators are shown in (Glazunov, et al. 1999) which were invented by (Kraynev, & Glazunov, 1991). Such mechanisms may be utilized to manipulate the corrosive medium at all actuators that are located out of the working space. Existence of several sub-chains and many closed loops determine the essential complication of the mathematical description of these mechanisms. Screw calculus using screw groups is universal and effective for parallel mechanisms analysis. Here, 1 describes the fixed base, 2 describes the output link and 3 describes the actuators. Addition, A_i expresses the spherical joint center situated on the fixed base; B_j expresses the center of the spherical joints combined with translational; C_j expresses the output link spherical joint centers; l_i, d_j expresses the generalized coordinates and s_j, f_j expresses the link lengths ($i=1,...,6$; $j=1,..., 3$).

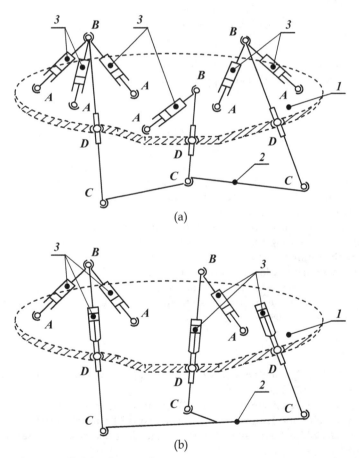

(a)

(b)

Fig. 4. The six-DOFs parallel mechanisms

In general, the wrench axis corresponding to i-th stalled actuator is located in the plane $(A_iB_jC_j)$, passes through center joint C_j and is directed perpendicular to its possible displacement. With the mechanisms as in (Fig. 3, a) the components of the wrench R_i can be find as:

$$r_i = \cfrac{1}{p_i \left\{ a_i + \left[\cfrac{s_j}{f_j} - \left(\cfrac{1}{f_j} + \cfrac{1}{s_j} \right) \cfrac{(a_i.b_j)}{d_j} \right] b_j \right\}} \; , \; r_i^0 = \rho_{C_j} \times r_i \tag{22}$$

where p_i - vector defining the wrench axis, a_i - vector from point A_i to point C_j, b_i - vector from point D_j to point C_j and ρ_{C_j} - radius vector point C_j.

Besides, with the mechanism as in (Fig. 4, b) the wrench axis $(j=1, \ldots,3)$ coincides with actuator axis. The wrench of the i-th stalled actuator $(i=, \ldots,3)$ is given as

$$r_i = \frac{1}{l_i} \left\{ a_i - \frac{(a_i.b_j) b_j}{d_j^2} \right\} , \; ; \; r_i^0 = \rho_{C_j} \times r_i \tag{23}$$

The present approach may be applied for different types of mechanisms such as sub-chains with varied actuator connection using spherical pairs.

8. Conclusion

Thus in this paper various criteria of design and singularity analysis of parallel manipulators are presented. The constraint wrenches imposed to the platform by kinematic chains is proposed to rely on the screw theory by determinants of matrix consisting of the Plücker coordinates of the unit screws. Criteria for design and singularity analysis of parallel manipulators with linear actuators located on the base are presented. The kinematic criterion of singularity corresponds to linear dependence of wrenches supporting the mobile platform; the static criterion corresponds to the limitation of pressure angles. The dynamical decoupling allows increasing the accuracy, parametrical optimization allows designing the mechanisms with optimal working volume, dexterity and stiffness, determination of the twists inside singularity allows finding the differential conditions of singular loci. Furthermore, the use of screw groups in order to determination of the singular zones of the multi-DOFs parallel mechanisms that make form of continuous areas and manipulators for external conditions are expressed.

9. References

Angeles, J. (2004). The Qualitative Synthesis of Parallel Mechanisms, In *Journal of Mechanical Design*, 126: 617-624.

Arakelian, V.; Briot, S. & Glazunov, V. (2007). Improvement of functional performance of spatial parallel mechanisms using mechanisms of variable structure, In *Proceedings of the 12th World Congress in Mechanism and Machine Science*, Besancon, France, 1: 159-164.

Bonev, I.; Zlatanov, D. & Gosselin, C. (2003). Singularity analysis of 3-DOF planar parallel mechanisms via screw theory, In *Transactions of the ASME, Journal of Mechanical Design*, 125: 573-581.

Ceccarelli, M. (2004). Fundamentals of Mechanics of Robotic Manipulations, *Kluwer Academic Publishers*.

Dimentberg, F. (1965). The Screw calculus and its Applications in Mechanics, *Nauka*, (English translation: AD680993, Clearinghouse for Federal Technical and Scientific Information, Virginia).

Glazunov, V. & Kraynev, A. (2006). Design and Singularity Criteria of Parallel mechanisms, *In ROMANSY 16, Robot Design, Dynamics, and Control, Proceedings of 16 CISM-IFToMM Symposium*, Springer Wien New York, 15-22.

Glazunov, V. & Thanh, N.M. (2008). Determination of the parameters and the Twists inside Singularity of the parallel Manipulators with Actuators Situated on the Base, *ROMANSY 17, Robot Design, Dynamics, and Control. In Proceedings of the Seventeenth CISM-IFToMM Symposium*, Tokyo, Japan, 467-474.

Glazunov, V. (2006). Twists of Movements of Parallel Mechanisms inside Their Singularities, *In Mechanism and Machine Theory*, 41: 1185-1195.

Glazunov, V.; Gruntovich, R.; Lastochkin, A. & Thanh, N.M. (2007). Representations of constraints imposed by kinematic chains of parallel mechanisms, *In Proceedings of 12th World Congress in Mechanism and Machine Science*, Besancon, France, 1: 380-385.

Glazunov, V.; Hue, N.N. & Thanh, N.M. (2009). Singular configuration analysis of the parallel mechanisms, *In Journal of Machinery and engineering education*, ISSN 1815-1051, No. 4, 11-16.

Glazunov, V.; Koliskor, A. & Kraynev, A. (1991). Spatial Parallel Structure Mechanisms, *Nauka*.

Glazunov, V.; Koliskor, A.; Kraynev, A. & B. Model, (1990). Classification Principles and Analysis Methods for Parallel-Structure Spatial Mechanisms, *In Journal of Machinery Manufacture and Reliability*, Allerton Press Inc., 1: 30-37.

Glazunov, V.; Kraynev, A.; Rashoyan, G. & Trifonova, A. (1999). Singular Zones of Parallel Structure Mechanisms, *In X World Congress on TMM*, Oulu, Finland, 2710-2715.

Gosselin, C. & Angeles, J. (1990). Singularity Analysis of Closed Loop Kinematic Chains, *In IEEE Trans. on Robotics and Automation*, 6(3): 281-290.

Huang, Z. (2004). The kinematics and type synthesis of lower-mobility parallel robot manipulators, *Proceedings of the XI World Congress in Mechanism and Machine Science*, Tianjin, China, 65–76.

Kong, X. & Gosselin, C. (2007). Type Synthesis of Parallel Mechanisms, *Springer-Verlag Berlin Heidelberg*, 272p.

Kraynev, A. & Glazunov, V. (1991). Parallel Structure Mechanisms in Robotics, *In MERO'91, Sympos. Nation. de Robotic Industr.*, Bucuresti, Romania, 1: 104-111.

Merlet, J.-P. (1991). Articulated device, for use in particular in robotics, *United States Patent 5,053,687*.

Merlet, J.-P. (2006). Parallel Robots, *Kluwer Academic Publishers*, 394p.

Statnikov, R.B. (1999). Multicriteria Design. Optimization and Identification, Dordrecht, Boston, London: *Kluwer Academic Publishers*, 206p.

Thanh, N.M.; Glazunov, V. & Vinh, L.N. (2010). Determination of Constraint Wrenches and Design of Parallel Mechanisms, *In CCE 2010 Proceedings, Tuxtla Gutiérrez, Mexico, 2010, International Conference on Electrical Engineering, Computing Science and Automatic Control, IEEE 2010*, 46-53.

Thanh, N.M.; Glazunov, V.; Tuan, T.C. & Vinh, N.X. (2010). Multi-criteria optimization of the parallel mechanism with actuators located outside working space, *In ICARCV 2010 Proceedings, Singapore, International Conference on Control, Automation, Robotics and Vision, IEEE 2010*, 1772-1778.

Thanh, N.M.; Glazunov, V.; Vinh, L.N. & Mau, N.C. (2008). Parametrical optimization of the parallel mechanisms while taking into account singularities, *In ICARCV 2008 Proceedings, Hanoi, Vietnam, International Conference on Control, Automation, Robotics and Vision, IEEE 2008*, 1872-1877.

Thanh, N.M.; Quoc, L.H. & Glazunov, V. (2009). Constraints analysis, determination twists inside singularity and parametrical optimization of the parallel mechanisms by means the theory of screws, *In CCE 2009 Proceedings, Toluca, Mexico, International Conference on Electrical Engineering, Computing Science and Automatic Control, IEEE 2009*, 89-95.

9

Optimization of H4 Parallel Manipulator Using Genetic Algorithm

M. Falahian, H.M. Daniali* and S.M. Varedi

Babol University of Technology
Iran

1. Introduction

Parallel manipulators have the advantages of high stiffness and low inertia compared to serial ones (Merlet, 2006). Most pick-and-place operations, including picking, packing and palletizing tasks; require four-degree-of-freedom (DOF), i.e. three translations and one rotation around a vertical axis (Company et al, 2003). A new family of 4-DOF parallel manipulator being called H4 that could be useful for high-speed, pick-and-place applications is proposed by Pierrot and Company (Pierrot & Company, 1999). This manipulator offers 3-DOF in translation and 1-DOF in rotation about a given axis. The H4 manipulator is useful for high-speed handling in robotics and milling in machine tool industry since it is a fully-parallel mechanism with no passive chain which can provide high performance in terms of speed and acceleration (Wu et al, 2006). Its prototype, built in the Robotics Department of LIRMM, can reach 10g accelerations and velocities higher than 5 m/s (Robotics Department of LIRMM). Pierrot et al. proved the efficiency of H4 serving as a high-speed pick-and-place robot (Pierrot et al, 2006). Corradini et al. evaluated the 4-DOFs parallel manipulator stiffness by two methods and compared the results (Coradini & Fauroux, 2003). Renaud et al. presented the kinematic calibration of a H4 robot using a vision-based measuring device (Renaud et al, 2003). Tantawiroon et al. designed and analyzed a new family of H4 parallel robots (Tantawiroon & Sangveraphunsiri, 2003). Poignet et al. estimated dynamic parameters of H4 with interval analysis (Poignet et al, 2003).

Parallel manipulators suffer from smaller workspaces relative to their serial counterparts; therefore, many researchers addressed the optimization of their workspaces (Boudreau & Gosselin, 1999; Laribi et al, 2007). But optimization for such a purpose might lead to a manipulator with poor dexterity. To alleviate this drawback some others considered both performance indices and volume of workspace, simultaneously (Li & Xu, 2006; Xu & Li, 2006; Lara et al, 2010).

This chapter deals with an optimal design of H4 parallel manipulator aimed at milling and Rapid-Prototyping applications with three degrees of freedom in translation and one in rotation. The forward and inverse kinematics of the manipulator are solved. The forward kinematics analysis of H4 leads to a univariate polynomial of degree eight. The workspace of the manipulator is parameterized using several design parameters. Some geometric constraints are considered in the problem, as well. Because of nonlinear discontinuous behaviour of the

---

* Corresponding Author

problem, Genetic Algorithm (GA) Method is used here to optimize the workspace. Finally, using GA, the manipulator is optimized based on a mixed performance index that is a weighted sum of global conditioning index and its workspace. It is shown that by introducing this measure, the parallel manipulator is improved at the cost of workspace reduction.

2. Description and mobility analysis of a H4 parallel manipulator

The basic concept of H4 is described by a simple architectural scheme as illustrated in Fig. 1, where joints are represented by lines (Pierrot et al, 2001). The manipulator is based on four independent chains between the base and the End-Effector (EE); each chain is driven by an actuator which is fixed on the base. Let P, R, U and S represent prismatic, revolute, universal, and spherical joints, respectively. Each of the P-U-U and R-U-U chains must satisfy some geometrical conditions to guarantee that the manipulator offers three translations and one rotation about a given axis (see (Pierrot & Company, 1999) for details). The U-U and $(S$-$S)_2$ (or $(U$-$S)_2$) chains can be considered as "equivalent" in terms of number and type of degrees of motion, so the P-U-U (or R-U-U) chains can be replaced by P-$(U$-$S)_2$ or P-$(S$-$S)_2$ chains (respectively by R-$(U$-$S)_2$ or R-$(S$-$S)_2$). In this chapter, P-$(S$-$S)_2$ chain is chosen for the mechanism architecture of H4. Fig. 2 shows the architecture of H4 (chain #1 is not plotted for sake of simplicity) (Poignet et al, 2010).

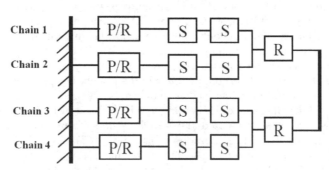

Fig. 1. Architectural scheme of H4

It is noteworthy that the four independent chains are not directly connected to the EE. It is possible to add another revolute joint on this end part and to couple it (with gears, timing belt, etc (Robotics Department of LIRMM)) so that the range of motion obtained on this latter joint can be larger than the end part ones. It is even possible to equip this passive revolute joint with a sensor to improve the robot accuracy. Moreover, the manipulator is a high-speed mechanism and can be easily controlled. So the H4 manipulator has the ability of serving as an efficient pick-and-place robot.

Considering the H4 robotic structure as given in Fig. 2, its geometrical parameters of are defined as follows:

- Two frames are defined, namely {A}: a reference frame fixed on the base; {B}: a coordinate frame fixed on the EE (C_1-C_1).
- The actuators slide along guide-ways oriented along a unitary vector, $\vec{k}_z$ ($\vec{k}_z$ is the unity vector parallel to the z axis in the reference frame {A}), and the origin is point $\mathbf{P}_i$, so the position of each point $\mathbf{A}_i$ is given by: $\vec{\mathbf{A}}_i = \vec{\mathbf{P}}_i + \mathbf{q}_i \vec{\mathbf{z}}_i$; for i=1,...4, in which $\mathbf{q}_i$ is the actuator coordinates.

- The parameters M_i, d and h are the length of the rods, the offset of the revolute-joint from the ball-joint, and the offset of each ball-joint from the center of the traveling plate, respectively.
- The pose of the EE is defined by a position vector $\vec{B} = \begin{bmatrix} x & y & z \end{bmatrix}^T$ and an angle θ, representing its orientation.

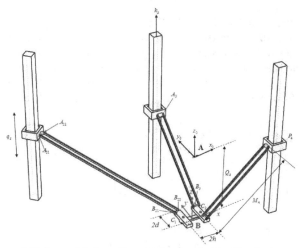

a) Robot with four lines drives and $(S-S)_2$ chains

b) Robot CAD model

Fig. 2. H4 parallel manipulator

Without losing generality, in this chapter we consider $\vec{P_i} = \begin{bmatrix} a_i & b_i & 0 \end{bmatrix}^T$ for simplicity, Q_z is the offset of {B} from {A} along the direction of $\vec{k_z}$.

One can effectively analyze the mobility of a H4 parallel manipulator by resorting to screw theory, which is a convenient tool to study instantaneous motion systems that include both rotation and translation in three dimensional spaces (Zhao et al, 2004).

A screw is called a twist when it is used to describe the motion state of a rigid body and a wrench when it is used to represent the force and moment of a rigid body. If a wrench acts on a rigid body in such a way that it produces no work, while the body is undergoing an infinitesimal twist, the two screws are said to be reciprocal (Li & Xu, 2007). In a parallel manipulator, the reciprocal screws associated with a serial limb represent the wrench constraints imposed on the moving platform by the serial limb. The motions of the moving platform are determined by the combined effect of all the wrench constraints imposed by each limb. Therefore, we can get the following equation:

$$\$_r^T \circ \$ = 0 \tag{1}$$

where $\$_r$ is the reciprocal screw and "$\circ$" represents the reciprocal production of two screws (Dai & Jones, 2007). According to the physical interpretations of reciprocal screws, one can obtain the constraints spaces that should be spanned by all of the reciprocal screws. As far as a H4 parallel manipulator is concerned, each S joint is equivalent to three intersecting non-coplanar R joints, and then the joint twists associated with the ith $P\text{-}(S\text{-}S)_2\text{-}R$ (or $P\text{-}U\text{-}U\text{-}R$) chain form a 6-system, which can be identified in the fixed frame as follows:

$$\$_1^i = \begin{bmatrix} 0 \\ \vec{s}_1^i \end{bmatrix}, \qquad \$_2^i = \begin{bmatrix} {}_1\vec{s}_{A_i}^i \\ A_i \times {}_1\vec{s}_{A_i}^i \end{bmatrix}, \qquad \$_3^i = \begin{bmatrix} {}_2\vec{s}_{A_i}^i \\ A_i \times {}_2\vec{s}_{A_i}^i \end{bmatrix}$$

$$\$_4^i = \begin{bmatrix} {}_1\vec{s}_{B_i}^i \\ B_i \times {}_1\vec{s}_{B_i}^i \end{bmatrix}, \qquad \$_5^i = \begin{bmatrix} {}_2\vec{s}_{B_i}^i \\ B_i \times {}_2\vec{s}_{B_i}^i \end{bmatrix}, \qquad \$_6^i = \begin{bmatrix} \vec{s}_{C_i}^i \\ C_i \times \vec{s}_{C_i}^i \end{bmatrix} \tag{2}$$

where $\vec{s}_j^i$ denotes a unit vector along the jth joint axis of the ith chain, for $i = 1,...,4$. Therefore, the kinematic screws of each kinematic chain can be expressed as:

$$[\$^i]^T = \begin{bmatrix} \$_1^i & \$_2^i & \$_3^i & \$_4^i & \$_5^i & \$_6^i \end{bmatrix} \tag{3}$$

Therefore, the reciprocal screws of kinematic chains can be derived as:

$$[\$_{platform}^r]^T = \begin{bmatrix} \$_r^1 & \$_r^2 & \$_r^3 & \$_r^4 \end{bmatrix} \tag{4}$$

Finally one can compute null spaces of the foregoing reciprocal screws which leads to the DOF of the manipulator; namely,

$$\begin{aligned}
DoF_1 &= \begin{bmatrix} 0 & 0 & 0 & 1 & 0 & 0 \end{bmatrix}^T \\
DoF_2 &= \begin{bmatrix} 0 & 0 & 0 & 0 & 1 & 0 \end{bmatrix}^T \\
DoF_3 &= \begin{bmatrix} 0 & 0 & 0 & 0 & 0 & 1 \end{bmatrix}^T \\
DoF_4 &= \begin{bmatrix} 0 & 0 & 1 & 0 & 0 & 0 \end{bmatrix}^T
\end{aligned} \tag{5}$$

These imply that H4 robot has three translational and one rotational DOF. Figure 3 shows the rotation of C_1C_2 link about z-axis as rotational DOF.

3. Kinematic analysis

3.1 Inverse kinematics

The inverse position kinematics problem solves the actuated variables from a given pose of EE. Let $\mathbf{q} = \begin{bmatrix} q_1 & q_2 & q_3 & q_4 \end{bmatrix}^T$ be the array of the four actuator joint variables and

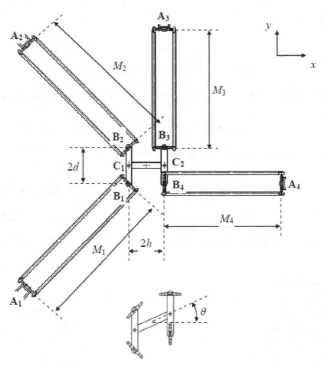

Fig. 3. Design parameters of H4

$\vec{x} = \begin{bmatrix} x & y & z & \theta \end{bmatrix}^T$ be the array of the pose of the EE. The position of points C_1 and C_2 are given by [4]:

$$\vec{C_1} = \vec{B} + \mathbf{Rot}(\theta)\overrightarrow{BC_1} \qquad \vec{C_2} = \vec{B} + \mathbf{Rot}(\theta)\overrightarrow{BC_2} \qquad (6)$$

where $\mathbf{Rot}(\theta)$ denotes the rotation matrix of the EE with respect to reference frame and is defined as:

$$\mathbf{Rot}(\theta) = {}^A\mathbf{R}_B = \mathbf{R}_z(\theta)$$
$$= \begin{bmatrix} cos(\theta) & -sin(\theta) & 0 \\ sin(\theta) & cos(\theta) & 0 \\ 0 & 0 & 1 \end{bmatrix} \qquad (7)$$

Moreover, the position of each point B_i is given by:

$$\vec{B_1} = \vec{C_1} + \overrightarrow{C_1B_1} \qquad \vec{B_2} = \vec{C_1} + \overrightarrow{C_1B_2}$$
$$\vec{B_3} = \vec{C_2} + \overrightarrow{C_2B_3} \qquad \vec{B_4} = \vec{C_2} + \overrightarrow{C_2B_4} \qquad (8)$$

Therefore, the following can then be written:

$$||A_iB_i||^2 = M_i^2 \qquad (9)$$

Then, for the first chain:

$$\overrightarrow{A_1B_1} = (\vec{B} + \text{Rot}(\theta)\overrightarrow{BC_1} + \overrightarrow{C_1B_1} - \vec{P}_1) - q_1\vec{z}_1$$
$$(A_1B_1)^2 = q_1^2 - 2q_1d_1.z_1 + \|d_1\|^2 \tag{10}$$

where $\vec{d}_1 = \vec{B} + \text{Rot}(\theta)\overrightarrow{BC_1} + \overrightarrow{C_1B_1} - \vec{P}_1$. Finally, the two solutions of Eq. 10 are given by:

$$q_1 = \vec{d}_1.\vec{z}_1 \pm \sqrt{(\vec{d}_1.\vec{z}_1)^2 + M_1^2 - \|d_1\|^2} \tag{11}$$

Similarly, q_1, q_2 and q_4 can be derived as:

$$q_2 = \vec{d}_2.\vec{z}_2 \pm \sqrt{(\vec{d}_2.\vec{z}_2)^2 + M_2^2 - \|d_2\|^2} \qquad \vec{d}_2 = \vec{B} + \text{Rot}(\theta)\overrightarrow{BC_1} + \overrightarrow{C_1B_2} - \vec{P}_2$$
$$q_3 = \vec{d}_3.\vec{z}_3 \pm \sqrt{(\vec{d}_3.\vec{z}_3)^2 + M_3^2 - \|d_3\|^2} \qquad \vec{d}_3 = \vec{B} + \text{Rot}(\theta)\overrightarrow{BC_2} + \overrightarrow{C_2B_3} - \vec{P}_3 \tag{12}$$
$$q_4 = \vec{d}_4.\vec{z}_4 \pm \sqrt{(\vec{d}_4.\vec{z}_4)^2 + M_4^2 - \|d_4\|^2} \qquad \vec{d}_4 = \vec{B} + \text{Rot}(\theta)\overrightarrow{BC_2} + \overrightarrow{C_2B_4} - \vec{P}_4$$

3.2 Forward kinematics
Forward kinematic problem solves the pose of the EE from a given actuators variables. It is well-known that forward kinematic problem of parallel manipulators is challenging (Dasgupta & Mruthyunjaya, 2000) and involves the solution of a system of nonlinear coupled algebraic equations in the variables describing the platform posture. This system of nonlinear equations might lead to solutions. Except in a limited number of the problems, one has difficulty in finding exact analytical solutions for the problem. So these nonlinear simultaneous equations should be solved using other methods, namely, numerical or semi-exact analytical methods. Some others believe that the combination of numerical and semi-exact analytical methods can also produce useful results (Hashemi et al, 2007; Varedi et al, 2009). Choi et al. solved forward kinematic problem of H4 and showed that the problem lead to a 16th degree polynomial in a single variable (Choi et al, 2003).
The EE is composed of three parts (two lateral bars and one central bar). Moreover, points B_1, B_2, B_3 and B_4 form a parallelogram. Let ^AA_i and ^AB_i represent the homogeneous coordinates of the points A_i in {A} and B_i in {A}, respectively. Then, one can write (Wu et al, 2006):

$$^A\vec{B}_i = \begin{bmatrix} x + hE_{1i}c\theta \\ y + hE_{1i}s\theta + dE_{2i} \\ z \\ 1 \end{bmatrix} = \begin{bmatrix} P_ic\theta + x \\ P_is\theta + y + Q_i \\ z \\ 1 \end{bmatrix} \tag{13}$$

$$^A\vec{A}_i = \begin{bmatrix} a_i & b_i & q_i & 1 \end{bmatrix}^T$$

where

$$P_i = hE_{1i}, \ Q_i = dE_{2i}, \ E_{11} = E_{12} = -1, \ E_{13} = E_{14} = -1,$$
$$E_{21} = E_{24} = 1, \ E_{22} = E_{23} = 1, \ c\theta = \cos(\theta), \ s\theta = \sin(\theta) \tag{14}$$

Substituting these values in Eq.(9), upon simplifications, yields:

$$(P_i\, c\theta + x - a_i)^2 + (P_i\, s\theta + y + Q_i - b_i)^2 + (z - q_i)^2 = M_i^2 \quad i = 1,...,4 \tag{15}$$

Expanding and rearranging Eq.(15) leads to:

$$\tilde{A}_i\, x + \tilde{B}_i\, y + \tilde{C}_i\, z + \tilde{D}_i + \tilde{E}_i = 0 \quad for \ i = 1,...,4$$

$$\tilde{A}_i = 2(P_i\, c\theta - a_i), \quad \tilde{B}_i = 2(P_i\, s\theta + Q_i - b_i) \quad \tilde{C}_i = -2q_i$$

$$\tilde{D}_i = a_i^2 + b_i^2 + q_i^2 - 2P_i(a_i\, c\theta + b_i\, s\theta - Q_i\, s\theta) - 2Q_i\, b_i + P_i^2 + Q_i^2 - M_i^2 \tag{16}$$

$$\tilde{E}_i = x^2 + y^2 + z^2$$

Subtracting Eq.(16) for i=3, 4 from the same equation for i=2, yields to:

$$\Delta A_{23}\, x + \Delta B_{23}\, y + \Delta C_{23}\, z + \Delta D_{23} = 0 \tag{17}$$

$$\Delta A_{24}\, x + \Delta B_{24}\, y + \Delta C_{24}\, z + \Delta D_{24} = 0 \tag{18}$$

where

$$\Delta A_{23} = (\tilde{A}_2 - \tilde{A}_3), \Delta B_{23} = (\tilde{B}_2 - \tilde{B}_3), \Delta C_{23} = (\tilde{C}_2 - \tilde{C}_3), \Delta D_{23} = (\tilde{D}_2 - \tilde{D}_3),$$

$$\Delta A_{24} = (\tilde{A}_2 - \tilde{A}_4), \Delta B_{24} = (\tilde{B}_2 - \tilde{B}_4), \Delta C_{24} = (\tilde{C}_2 - \tilde{C}_4), \Delta D_{24} = (\tilde{D}_2 - \tilde{D}_4),$$

Eqs. (17) and (18) now can be solved for x and y, namely:

$$x = \frac{\Delta_{11} z + \Delta_{12}}{\Delta_0} = e_1 z + e_2 \tag{19}$$

$$y = \frac{\Delta_{21} z + \Delta_{22}}{\Delta_0} = e_3 z + e_4 \tag{20}$$

where

$$e_1 = \frac{\Delta_{11}}{\Delta_0}, e_2 = \frac{\Delta_{12}}{\Delta_0}, e_3 = \frac{\Delta_{21}}{\Delta_0}, e_4 = \frac{\Delta_{22}}{\Delta_0}, \ \Delta_0 = \Delta A_{23}\Delta B_{24} - \Delta A_{24}\Delta B_{23},$$

$$\Delta_{11} = \Delta B_{23}\Delta C_{24} - \Delta B_{24}\Delta C_{23}, \ \Delta_{12} = \Delta B_{23}\Delta D_{24} - \Delta B_{24}\Delta D_{23},$$

$$\Delta_{21} = \Delta A_{24}\Delta C_{23} - \Delta A_{23}\Delta C_{24}, \ \Delta_{22} = \Delta A_{24}\Delta D_{23} - \Delta A_{23}\Delta D_{24},$$

Substituting Eqs.(19) and (20) into Eq.(16) for i=2, yields to the following quadratic equation.

$$\lambda_0 z^2 + \lambda_1 z + \lambda_2 = 0 \tag{21}$$

where

$$\lambda_0 = e_1^2 + e_3^2 + 1, \quad \lambda_1 = 2e_1 e_2 + 2e_3 e_4 + \tilde{A}_2 e_1 + \tilde{B}_2 e_3 + \tilde{C}_2,$$

$$\lambda_2 = e_2^2 + e_4^2 + \tilde{A}_2 e_2 + \tilde{B}_2 e_4 + \tilde{D}_2$$

Eq.(21) can be solved for z. Substituting this value into Eq.(16) for i=1, upon simplifications leads:

$$f((c\theta)^4, (c\theta)^3 (s\theta)^1, (c\theta)^2 (s\theta)^2, (c\theta)^1 (s\theta)^3, (c\theta)^0 (s\theta)^4, (c\theta)^3 (s\theta)^0,$$

$$(c\theta)^2 (s\theta)^1, (c\theta)^1 (s\theta)^2, (c\theta)^0 (s\theta)^3, (c\theta)^2 (s\theta)^0, (c\theta)^1 (s\theta)^1, (c\theta)^0 (s\theta)^2, \tag{22}$$

$$(c\theta)^1 (s\theta)^0, (c\theta)^0 (s\theta)^1, (c\theta)^0 (s\theta)^0) = 0$$

where $f(x_1, x_2, ..., x_n)$ represents a linear combination of $x_1, x_2, ..., x_n$. The coefficients of Eq.(22) are the data depend on the design parameters of the H4 robot. In Eq.(22), we substitute now the equivalent expressions for $c\theta$ and $s\theta$ given as:

$$c\theta = \frac{1-T^2}{1+T^2} \quad , \quad s\theta = \frac{2T}{1+T^2} \quad , \quad T = tan(\theta/2) \tag{23}$$

Upon simplifications, Eq.(22) leads to a univariate polynomial of degree eight, namely;

$$aT^8 + bT^7 + cT^6 + dT^5 + eT^4 + fT^3 + gT^2 + hT^1 + i = 0 \tag{24}$$

where a, b, c, d, e, f, g, h and i depend on kinematic parameters. The detailed expressions for them are not given here because these expansions would be too large to serve any useful purpose. What is important to point out here is that the above equation admits eight solutions, whether real or complex, among which we only interested in the real ones. Substituting the values of T from Eq.(23) into Eq.(19), Eq.(20) and Eq.(21), complete the solutions.

3.2.1 Numerical example of kinematic analysis
Here, we include a numerical example as shown in Fig.4. The design parameters of H4 are given as:

$$h = d = 6(cm), \ Q_z = 42(cm), \ M_1 = M_2 = \sqrt{3}Q_z, \ M_3 = M_4 = \sqrt{2}Q_z \tag{25}$$

Substituting these values in Eq.(13), yields:

$$\vec{A}_1 = [-48 \ \ -48 \ \ 0]^T, \ \vec{B}_1 = [-6 \ \ -6 \ \ -42]^T,$$
$$\vec{A}_2 = [-48 \ \ 48 \ \ 0]^T, \ \vec{B}_2 = [-6 \ \ 6 \ \ -42]^T,$$
$$\vec{A}_3 = [6 \ \ 48 \ \ 0]^T, \ \ \ \ \vec{B}_3 = [6 \ \ 6 \ \ -42]^T,$$
$$\vec{A}_4 = [48 \ \ -6 \ \ 0]^T, \ \vec{B}_4 = [6 \ \ -6 \ \ -42]^T.$$

Given, the position and orientation of the EE as $\vec{x} = [5 \ \ -5 \ \ -30 \ \ \pi/6]^T$, inverse kinematics are given by:

$$\vec{q} = [13.021 \ \ -7.488 \ \ 9.679 \ \ 15.77]^T \tag{26}$$

Next, the forward kinematic problem, for the actuators coordinate as given in Eq.(26), is solved. The polynomial equation in Eq.(24) can be solved for T, namely;

$$\begin{aligned}
T_{12} &= -0.61309 \pm 0.56609i \\
T_{34} &= 0.073172 \pm 0.84245i \\
T_{56} &= 0.028294 \pm 0.88575i \\
T_7 &= 0.267957 \\
T_8 &= -2.8822
\end{aligned} \tag{27}$$

There are only two real solutions, for which $T_7 = 0.2680$ leads to $\vec{x} = [5(cm) \ \ -5(cm) \ \ -30(cm) \ \ 0.524(rad)]^T$ and $T_8 = -2.8822$, yields:

$\vec{x} = \begin{bmatrix} 3(cm) & -7.95(cm) & -14.58(cm) & -2.47(rad) \end{bmatrix}^{T}$. However, the latter configuration cannot be realized in practice because of the mutual interference of the parallelograms as shown in Fig.4.

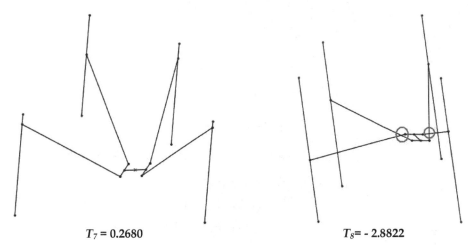

$T_7 = 0.2680$ $\qquad\qquad\qquad\qquad$ $T_8 = -2.8822$

Fig. 4. The solutions for forward kinematic problem

3.3 Jacobian matrix

Jacobian matrix relates the actuated joint velocities to the EE Cartesian velocities, and is essential for the velocity and trajectory control of parallel robots. The velocity of the EE of H4 can be defined by resorting to a velocity for the translation, $\vec{v}_B = \begin{bmatrix} \dot{x} & \dot{y} & \dot{z} \end{bmatrix}^{T}$ and a scalar for the rotation about $\vec{z}$; namely, $\dot{\theta}$. Thus, the velocity of points C_1 and C_2 can be written as follows (Wu et al, 2006):

$$\vec{v}_{C_1} = \vec{v}_B + \dot{\theta}(\vec{k}) \times \overrightarrow{BC_1} \qquad \vec{v}_{C_2} = \vec{v}_B + \dot{\theta}(\vec{k}) \times \overrightarrow{BC_2} \qquad (28)$$

Moreover, since the links B_1B_2 and B_3B_4 only have translational motion, the following relations hold:

$$\vec{v}_{B_1} = \vec{v}_{B_2} = \vec{v}_{C_1} \qquad \vec{v}_{B_3} = \vec{v}_{B_4} = \vec{v}_{C_2} \qquad (29)$$

On the other hand, velocity of points A_i is given by:

$$\vec{v}_{A_i} = \dot{q}\overrightarrow{k_z} \qquad (30)$$

The velocity relationship can then be written thanks to the classical property; namely,

$$\vec{v}_{A_i} \cdot \overrightarrow{A_iB_i} = \vec{v}_{B_i} \cdot \overrightarrow{A_iB_i} \qquad (31)$$

Eq.(31) can be written for i=1,...,4 and the results grouped in a matrix form as:

$$J_q\dot{\vec{q}} = J_x\dot{\vec{x}} \qquad (32)$$

Where $\dot{\mathbf{x}}$ and $\dot{\mathbf{q}}$ denote the vectors of the EE velocity and input actuated joint rates, respectively ; J_q and J_x are the Jacobian matrices, all are defined as:

$$\dot{\mathbf{q}} = \begin{bmatrix} \dot{q}_1 & \dot{q}_2 & \dot{q}_3 & \dot{q}_4 \end{bmatrix}^T$$

$$J_q = \begin{bmatrix} \overrightarrow{A_1B_1} \cdot \mathbf{z}_1 & 0 & 0 & 0 \\ 0 & \overrightarrow{A_2B_2} \cdot \mathbf{z}_2 & 0 & 0 \\ 0 & 0 & \overrightarrow{A_3B_3} \cdot \mathbf{z}_3 & 0 \\ 0 & 0 & 0 & \overrightarrow{A_4B_4} \cdot \mathbf{z}_4 \end{bmatrix}$$

$$\dot{\mathbf{x}} = \begin{bmatrix} \dot{x} & \dot{y} & \dot{z} & \dot{\theta} \end{bmatrix}^T \tag{33}$$

$$J_x = \begin{bmatrix} (\overrightarrow{A_1B_1})_x & (\overrightarrow{A_1B_1})_y & (\overrightarrow{A_1B_1})_z & (\overrightarrow{A_1B_1} \times \overrightarrow{BC_1}).\vec{\mathbf{k}} \\ (\overrightarrow{A_2B_2})_x & (\overrightarrow{A_2B_2})_y & (\overrightarrow{A_2B_2})_z & (\overrightarrow{A_2B_2} \times \overrightarrow{BC_1}).\vec{\mathbf{k}} \\ (\overrightarrow{A_3B_3})_x & (\overrightarrow{A_3B_3})_y & (\overrightarrow{A_3B_3})_z & (\overrightarrow{A_3B_3} \times \overrightarrow{BC_2}).\vec{\mathbf{k}} \\ (\overrightarrow{A_4B_4})_x & (\overrightarrow{A_4B_4})_y & (\overrightarrow{A_4B_4})_z & (\overrightarrow{A_4B_4} \times \overrightarrow{BC_2}).\vec{\mathbf{k}} \end{bmatrix}$$

If the mechanism is not in a singular configuration, the Jacobian and its inverse are described as:

$$J = J_x^{-1} J_q \quad and \quad J^{-1} = J_q^{-1} J_x \tag{34}$$

4. Optimization of H4

4.1 Workspace evaluation

For a robot in the context of industrial application and given parameters, it is very important to analyze the volume and the shape of its workspace (Dombre & Khalil, 2007). Calculation of the workspace and its boundaries with perfect precision is crucial, because they influence the dimensional design, the manipulator's positioning in the work environment, and its dexterity to execute tasks (Li & Xu, 2006). The workspace is limited by several conditions. The prime limitation is the boundary obtained through solving the inverse kinematic problem. Further, the workspace is limited by the reachable extent of drives and joints, then by the occurrence of singularities, and finally by the links and platform collisions (Stan et al, 2008). Various approaches may be used to calculate the workspace of a parallel robot. In the present work the boundary of workspace of H4 is determined geometrically (Merlet, 2006).

4.2 Condition number

In order to guarantee the regular workspace to be effective, constraints on the dexterity index are introduced into the design problem. A frequently used measure for dexterity of a manipulator is the inverse of the condition number of the kinematics Jacobian matrix (Hosseini et al, 2011). Indeed, in order to determine the condition number of the Jacobian matrices, their singular values must be ordered from largest to smallest. However, in the presence of positioning and orienting tasks the singular values associated with positioning, are dimensionless, while those associated with orientation have units of length, thereby making impossible such an ordering. This dimensional in-homogeneity can be resolved by introducing a weighting factor/length, which is here equal to the length of the first link (Hosseini et al, 2011). Then, the condition number of homogeneous Jacobian is defined as the ratio of its largest singular value σ_l to its smallest one, σ_s (Gosselin & Angeles, 1991), i.e.

$$\kappa(J) \equiv \frac{\sigma_l}{\sigma_s} \qquad (35)$$

Note that $\kappa(J)$ can attain any value from 1 to infinity, in which they correspond to isotropy and singularity, respectively.

4.3 Performance index

The objective function for maximization is defined here as a mixed performance index which is a weighted sum of global dexterity index (GDI), η_d and space utility ratio index (SURI), η_s (Dasgupta & Mruthyunjaya, 2000) , i.e.,

$$\eta = w_d \eta_d + (1 - w_d) \eta_s$$
$$= w_d \left[\frac{1}{V} \int_V \frac{1}{\kappa} dV \right] + (1 - w_d) \frac{V}{V^*} \qquad (36)$$

where the weight parameter $w_d (0 \le w_d \le 1)$ describes the proportion of GDI in the mixed index, and V^* represents the robot size which is evaluated by the product of area of the fixed platform and the maximum reach of the EE in the z direction.

4.4 Design variables

The architectural parameters of a H4 involve the offset of {B} from {A} along the direction of $\vec{k}_z(Q_z)$, length of the offset of the revolute-joint from the ball-joint (d) and the offset of each ball-joint from the center of the traveling plate (h). In order to perform the optimization independent of the dimension of each design candidate, the design variables are scaled by Δq, i.e., the motion range of the actuators. Thus, the design variables become

$\vec{\Gamma} = \left[\dfrac{d}{\Delta q}, \dfrac{h}{\Delta q}, \dfrac{Q_z}{\Delta q} \right]$. However, we have the following limitations for the design variables:

$$1 / \kappa(J) > 0.01,$$

$$0.01 < \frac{d}{\Delta q} < 0.15 \quad , \quad 0.01 < \frac{h}{\Delta q} < 0.15 \quad , \quad 0.35 < \frac{Q_z}{\Delta q} < 0.55$$

4.5 Optimization result of H4

In order to perform an optimal design of H4 parallel robots, an objective function was developed first, then GA is applied in order to optimize the objective function (Xu & Li, 2006). For optimization settings, regarding to the GA approach, normalized geometric selection is adopted (Merlet, 2006), and the genetic operators are chosen to be non-uniform mutation with the ratio of 0.08 and arithmetic crossover with the ratio of 0.8. Moreover, the population size is 30 and the generations' number is set to 150.

4.5.1 Optimization based on maximum workspace volume

First we consider the volume of workspace as the objective function; namely,

$V^* = max(V)$

Subject to

1. $1/\kappa(J) > 0.01$

$$0.01 < \frac{d}{\Delta q} < 0.15 \quad , \quad 0.01 < \frac{h}{\Delta q} < 0.15 \quad , \quad 0.35 < \frac{Q_z}{\Delta q} < 0.55$$

The solution of this problem is given in Table 1. Moreover, the workspace for $\theta = 0$ and the conditioning index for $\theta = 0$ and $z=0$ are shown in Fig.5 and Fig.6, respectively. While the workspace size is acceptable, the manipulator suffers from a poor dexterity throughout its workspace.

Volume of workspace *(rad)(cm³)*	Average of conditioning index	Robot's parameters		
		d(cm)	*h(cm)*	*Q_z (cm)*
12582.0	0.042	13.52484	13.88781	35.18892

Table 1. Optimization result for H4

As it is observed the GA reaches a convergence after 30 iterations as depicted in Fig. 7.

4.5.2 Optimization based on GDI

In this section we optimize the average of conditioning index in the workspace. Therefore, the problem is stated as:

$GDI = max(average\ (1/\ \kappa(J)))$

Subject to

1. $1/\kappa(J) > 0.01$

2. $0.01 < \frac{d}{\Delta q} < 0.15 \quad , \quad 0.01 < \frac{h}{\Delta q} < 0.15 \quad , \quad 0.35 < \frac{Q_z}{\Delta q} < 0.55$

As it is expected, the volume of the workspace is very small and only includes a small neighbourhood of a point of the best conditioning index. This optimal design is given in Table 2.

Volume of workspace *(rad)(cm³)*	Average of conditioning index	Robot's parameters		
		d(cm)	*h(cm)*	*Q_z (cm)*
4.1887	0.0705	1.00008	1.09692	35.48742

Table 2. Optimization result for H4

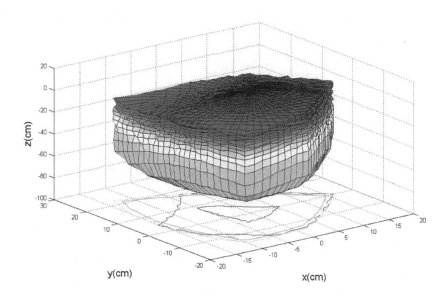

Fig. 5. Workspace of H4 for $\theta = 0$

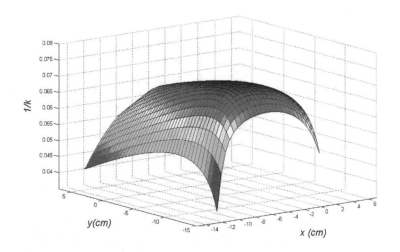

Fig. 6. Conditioning index for z=0 and $\theta = 0$

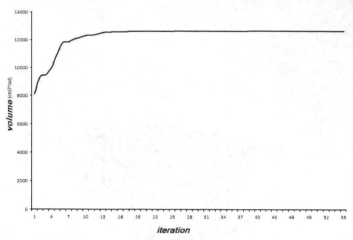

iteration

Fig. 7. Convergence of GA

4.5.3 Optimization based on the mixed performance index

For the problem at hand one needs a good fitness function, because the optimization based on maximal workspace volume, decreases the performance indices and the optimization based on GDI decreases the workspace volume. Therefore, one can optimize the manipulator based on performance index that described in subsection 5.3. The optimization results for w_d=0.25, 0.5, 0.75 are given in Table 3.

Moreover, for different values of w_d, the problem is solved and the GDI, SURI and the mixed performance index are calculated and plotted in Fig. 8. As the result, any value of w_d greater than 0.74 leads to a limited workspace and for any values smaller than that has no substantial effects on GDI and the workspace. Therefore, it clearly shows that by introducing this measure, the performance of the manipulator can be improved at a minor cost its workspace volume.

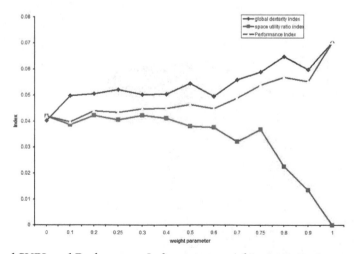

Fig. 8. GDI and SURI and Performance Index versus weight parameter

	$w_d = 0.25$	$w_d = 0.5$	$w_d = 0.75$
Volume of workspace *(rad)(cm³)*	11983	11304	11008
d(cm)	12.14243	12.60003	10.36257
h(cm)	10.38564	9.06147	5.56941
Q_z *(cm)*	37.25841	39.61105	40.69626
GDI	0.0521	0.0547	0.059

Table 3. Optimization results for H4

5. Conclusions

First, the forward and inverse kinematics of H4 parallel manipulator has been studied here, in which the former problem has leaded to a univariate polynomial of degree eight. Then, the optimal design of the manipulator has been addressed. Using genetic algorithm the manipulator has been optimized based on a mixed performance index that is a weighted sum of global conditioning index and its workspace volume. It has been shown that by introducing this measure, the parallel manipulator has been improved at minor cost of its workspace volume.

6. References

Boudreau, R.; Gosselin, C.M. (1999). The synthesis of planar parallel manipulators with a genetic algorithm, , *Journal of Mechanical Design*, 121(4), 1999, pp. 533–537.

Choi, H.B.; Konno, A. & Uchiyama, M. (2003). Closed-form solutions for the forward kinematics of a 4-DOFs parallel robot H4, *Proceedings of IEEE/RSJ International Conference on Intelligent Robots and Systems*, Las Vegas, USA, October, 2003, pp. 3312-3317.

Company, O.; Marquet, F. & Pierrot, F. (2003). A new high-speed 4DoF parallel robot synthesis and modeling issues. *IEEE, Transactions on Robotics and Automation*, Vol.19, No.3, 2003, pp. 411-420.

Coradini, C.; Fauroux, J. (2003). Evaluation of a 4-DOF parallel manipulator stiffness, *Mechanism and Machine Science*, Aug, 2003.

Dai, J.S.; Jones, J.R. (2007). A Linear Algebraic Procedure in Obtaining Reciprocal Screw Systems, *J. Robotic Systems*, 2007, 20(7), pp. 401–412.

Dasgupta, B.; Mruthyunjaya, T.S. (2000). The Stewart platform manipulator: A review, *Mechanism and Machine Theory*, 35, 2000, pp. 15–40.

Dombre, E.; Khalil, W. (2007). Modeling, Performance Analysis and Control of Robot Manipulators, *ISTE*, 2007.

Gosselin, C.; Angeles, J. (1991). A global performance index for the kinematic optimization of robotic manipulators. *ASME J Mech*, December, 1991, 113(3), pp. 220–226.

Hashemi, S.H.; Daniali, H.M. & Ganji, D.D. (2007). Numerical simulation of the generalized Huxley equation by He's homotopy perturbation method, *Applied Mathematics and Computation*, 2007, pp. 157–161.

Hosseini, M.A; Daniali, H.M. & Taghirad, H.D. (2011). "Dexterous Workspace Optimization of a Tricept Parallel Manipulator", *Advanced Robotics*, Vol. 25, 2011, pp. 1697-1712.

Lara-Molina, F.A.; Rosário, J.M. & Dumur, D. (2010). Multi-Objective Design of Parallel Manipulator Using Global Indices, *The Open Mechanical Engineering Journal*, 2010, 4, pp. 37-47.

Laribi, M.A.; Romdhane, L. & Zeghloul, S. (2007). Analysis and dimensional synthesis of the DELTA robot for a prescribed workspace, *Mechanism and Machine Theory*, 42, (2007), pp. 859–870.

Li, Yangmin.; Xu, Qingsong. (2006). A new approach to the architecture optimization of a general 3-PUU translational parallel manipulator. *J Int Robot System*, 2006, 46, pp. 59–72.

Li, Yangmin.; Xu, Qingsong. (2007). Kinematic analysis of a 3-PRS parallel manipulator, *Robotics and Computer-Integrated Manufacturing*, 2007, 23, pp. 395–408.

Merlet, J.P.; (2006). *Parallel robots* (second edition), springer.

Pierrot, F.; Company, O. (1999). H4: a new family of 4-DoF parallel robots, *Int Conf on Advanced Intelligent Mechatronics*, Proceedings of the 1999 IEEE/ASME, pp. 508-513.

Pierrot, F.; Marquet, F. & Company, O. & Gil, T. (2001). H4 parallel robot: modeling, design and preliminary experiments, *Int Conf on Robotics and Automation*, May, 2001.

Poignet, Ph.; Ramdani, N. & Vivas, O.A. (2003). Robust estimation of parallel robot dynamic parameters with interval analysis, *Proceedings of the 42nd IEEE*, Conference on Decision and Control, Maui, Hawaii USA, December, 2003.

Renaud, P.; Andreff, N. & Marquet, F. & Martinet, P. (2003). Vision-based kinematic calibration of a H4 parallel mechanism, *Int Conf on Robotics and Automation*, Proceediogr of the 2003 IEEE, Taipei, Taiwan, September 14-19.

Robotics Department of LIRMM. (n.d.). H4 with rotary drives: a member of H4 family, dedicated to high speed applications, http://www.lirmm.fr/w3rob.

Stan, S.D.; Manic, M. & Mătieş, M. & Bălan, R. (2008). Evolutionary approach to optimal design of 3 DOF Translation exoskeleton and medical parallel robots, *HSI 2008, IEEE Conference on Human System Interaction*, Krakow, Poland, May 25-27, 2008, pp. 720-725.

Tantawiroon, N.; Sangveraphunsiri, V. (2005). Design and analysis of a new H4 family parallel manipulator, *Thammasat Int J. Science Tech, July-September*, Vol. 10, No. 3, 2005.

Varedi, S.M.; Daniali, H.M. & Ganji, D.D. (2009). Kinematics of an offset 3-UPU translational parallel manipulator by the homotopy continuation method, *Nonlinear Analysis: Real World Applications 10*, 2009, pp. 1767–1774.

Wu, Jinbo.; Yin, Zhouping. & Xiong, Youlun. (2006). Singularity analysis of a novel 4-DoF parallel manipulator H4, *Int J Advanced Manuf Tech*, 29, pp. 794-802.

Xu, Qingsong.; Li, Yangmin. (2006). Kinematic analysis and optimization of a New compliant parallel micromanipulator, *Int J of Advanced Robotic Systems*, 2006, Vol. 3, No. 4.

Zhao, J.S.; Zhou, K. & Feng, Z.J. (2004). A theory of degrees of freedom for mechanisms. *Mechanism and Machine Theory*, 39, 2004, pp. 621–643.

10

Spatial Path Planning of Static Robots Using Configuration Space Metrics

Debanik Roy

Board of Research in Nuclear Sciences,
Department of Atomic Energy, Government of India, Mumbai
India

1. Introduction

Obstacle avoidance and robot path planning problems have gained sufficient research attention due to its indispensable application demand in manufacturing vis-à-vis material handling sector, such as picking-and- placing an object, loading / unloading a component to /from a machine or storage bins. Visibility map in the configuration space (*c-space*) has become reasonably instrumental towards solving robot path planning problems and it certainly edges out other techniques widely used in the field of motion planning of robots (e.g. Voronoi Diagram, Potential Field, Cellular Automata) for *unstructured environment*. The c-space mapping algorithms, referred in the paper, are discussed with logic behind their formulation and their effectiveness in solving path planning problems under various conditions imposed a-priori. The visibility graph (*v-graph*) based path planning algorithm generates the equations to obtain the desired joint parameter values of the robot corresponding to the ith intermediate location of the end - effector in the collision - free path. The developed c-space models have been verified by considering first a congested workspace in 2D and subsequently with the real spatial manifolds, cluttered with different objects. New lemma has been proposed for generating c-space maps for higher dimensional robots, e.g. having degrees-of-freedom more than three. A test case has been analyzed wherein a seven degrees-of-freedom revolute robot is used for articulation, followed by a case-study with a five degrees-of-freedom articulated manipulator (RHINO XR-3) amidst an in-door environment. Both the studies essentially involve new c-space mapping thematic in higher dimensions.

Tomas Lozano Perez' postulated the fundamentals of Configuration Space approach and proved those successfully in spatial path planning of robotic manipulators in an environment congested with polyhedral obstacles using an explicit representation of the manipulator configurations that would bring about a collision eventually [Perez', 1983]. However, his method suffers problem when applied to manipulators with revolute joints. In contrast to rectilinear objects, as tried by Perez', collision-avoidance algorithm in 2D for an articulated two-link planar manipulator with circular obstacles have been reported also [Keerthi & Selvaraj, 1989]. The paradigm of *automatic transformation* of obstacles in the c-space and thereby path planning is examined with finer details [De Pedro & Rosa, 1992], such as *friction* between the obstacles [Erdmann, 1994]. Novel c-space computation algorithm for convex planar algebraic objects has been reported [Kohler & Spreng, 1995], while *slicing* approach for the same is tried for curvilinear objects [Sacks & Bajaj, 1998] & [Sacks, 1999]. Nonetheless, various intricacies of the global c-space mapping techniques for a robot under static environment have been surveyed to a good extent [Wise & Bowyer, 2000]. Although the theoretical paradigms of c-space technique for solving *find-path* problem have been largely addressed in the above literature vis-à-vis a few more [Brooks, 1983], [Red & Truong-Cao, 1985], [Perez', 1987], [Hasegawa & Terasaki, 1988], [Curto & Monero, 1997], the bulging question of tackling collision detection under a typical manufacturing scenario, cluttered with real-life multi-featured obstacles remains largely unattended.

Survey reports on motion planning of robots in general, have been presented, with special reference to path planning problems of lower dimensionality [Schwartz & Sharir, 1988] & [Hwang & Ahuja, 1992]. The find-path problem under sufficiently cluttered environment has been studied with several customized models, such as using distance function [Gilbert & Johnson, 1985], probabilistic function [Jun & Shin, 1988], time-optimized function [Slotine & Yang, 1989], shape alteration paradigms [Lumelsky & Sun, 1990a] and sensorized stochastic method [Acar et al, 2003]. Even, novel *path transform function* for guiding the search for find-path in 2D is reported [Zelinsky, 1994], while the same for manipulators with higher degrees-of-freedom is also described [Ralli & Hirzinger, 1996]. All these treatises are appreciated from the context of theoretical estimation, but lacks in simulating all kinds of polyhedral obstacles.

Based on the c-space mapping, algorithmic path planning in 2D using *visibility* principle is studied [Fu & Liu, 1990], followed by exhaustive theoretical analysis on visibility maps [Campbell & Higgins, 1991]. However, issues regarding computational complexity involved in developing a typical visibility graph, which is O (n^2), 'n' being the total number of vertices in the map, is analyzed earlier [Welzl, 1985]. The concept of *M-line*[1] and its uniqueness in generating near-optimal solutions against heuristic-based search algorithms has also been examined [Lumelsky & Sun, 1990b].

Several researchers have reviewed the facets of path planning problem in a typical spatial manifold. A majority of these models are nothing but extrapolation of proven 2D techniques in 3D space [Khouri & Stelson, 1989], [Yu & Gupta, 2004] & [Sachs et al, 2004]. However, new methods for the generation of c-space in such cases (i.e. spatial) have been exploited too [Brost, 1989], [Bajaj & Kim, 1990] & [Verwer, 1990]. Customized solution for rapid computation of c-space obstacles has been addressed [Branicky & Newman, 1990], using geometric properties of collision detection between *known* static obstacles and the manipulator body, while sub-space method is being utilized in this regard [Red et al, 1987]. The usefulness of several new algorithms using v-graph technique has been demonstrated in spatial robotic workspace [Roy, 2005].

It may be mentioned at this juncture with reference to the citations above, that, although celebrated, a distinct methodology of using c-space mapping for higher dimensional robots as well as in spatial workspace is yet to be tuned. Our approach essentially calls on this lacuna of the earlier researches. We proclaim our novelty in adding new facets to the problem in a generic way, like: a] *rationalizing* configuration space mapping for *higher dimensional* (e.g. 7 or 8 degrees-of-freedom) *robots*; b] *preferential selection* of joint-variables for configuration space plots in 2D; c] extension of 2D path planning algorithm in 3D through *slicing technique* (creation, validation & assimilation of c-space slices) and d] *searching* collision-free path in 3D, using novel *visibility map-based algorithm*.

---

[1] Mean Line, as referred in the literature concerning the visibility graph-based path planning of robots.

The paper has been organized in six sections. The facets of our proposition towards configuration space maps in 3D are discussed in the next section. Section 3 delineates the c-space mapping algorithms, with the logistics and analytical models. The features of the path planning metrics and the algorithm in particular, using the concept of visibility graph, have been reported in section 4. Both 2D and spatial workspaces have been postulated, with an insight towards the analytical modeling, in respective cases alongwith test results. Section 5 presents the case study of robot path planning. Finally section 6 concludes the paper.

2. Configuration space map in 3D space: Our proposition

2.1 Modeling of robot workspace

The robotic environment is modeled through *discretization* of the 3D space into a number of 2D planes (Cartesian workspace), corresponding to a finite range of *waist* / base rotation of the robot[2]. Thus, modeling has been attempted with the sectional view of the obstacles in 2D plane.

The obstacles are considered to be *regular*, i.e. having finite shape and size with standard geometrical features with known vertices in Cartesian co-ordinates. For example, obstacles with shapes such as cube, rectangular paralleopiped, trapezoid, sphere, right circular cylinder, right pyramid etc. have been selected (as primitives) for modeling the environment. A complex obstacle has been modeled as a Boolean combination of these primitives, to have polygonal convex shape preferably. However, concave obstacles can also be used in the algorithms by approximating those to the nearest convex shapes, after considering their 'convex hulls' (polytones). Irregular-shaped obstacles have also been modeled by considering their envelopes to be of convex shapes. Circular obstacles have been approximated to the nearest squares circumscribing the original circles, thereby possessing *pseudo-vertices*.

Features of the developed technique, namely, "Slicing Method", are: i] alongwith shoulder, elbow and wrist (pitch only) rotations, waist rotation of the robot is considered, which is guided by a finite range vis-à-vis a finite resolution; ii] the entire 3D workspace is divided into a number of 2D planes, according to the number of 'segments' of the waist rotation; iii] for every fixed angle of rotation of the waist, a 2D plane is to be constructed, where all other variables like shoulder, elbow and wrist pitch movements are possible; iv] corresponding to each of the 2D slices of the workspace, either one obstacle entirely or a part of it will be generated, depending on the value of the resolution chosen for waist rotation.

Corresponding to each slice, one c-space map (considering only two joint variables at a time) is to be developed and likewise, several maps will be obtained for all the remaining slices. The final combination of the colliding joint variable values will be the union of all those sets of the values for each slice. Nevertheless, the process of computation can be simplified by taking some finite number of slices, e.g. four to five planes. Figure 1 pictorially illustrates the above-mentioned postulation, wherein the robotic workspace consists of various categories of obstacles, like regular geometry (e.g. obstacle 'A', 'B' & 'C') and integrated geometry[3] (e.g. obstacle 'D' & 'E').

---

[2] Base rotation is earmarked for articulated robots, whereas suitable angular divisions of the entire planar area, i.e. 360^0 are considered for robots with non-revolute joints (e.g. prismatic).

[3] Boolean combination of regular geometries is considered.

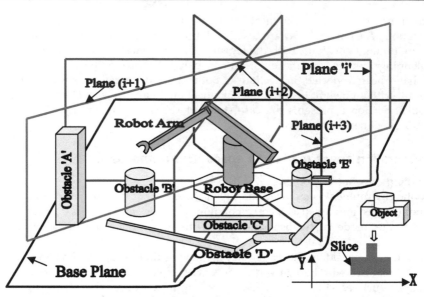

Fig. 1. Schematic view of 'Slicing Model' for a congested spatial robotic environment

By knowing the intercept co-ordinates for each slice, an equivalent 2D *slice workspace* is developed; simply by projecting those intercept lines. Obviously the height of the obstacle will now play a vital role and the projecting length will be equal to the height of the obstacle. In a similar manner, obstacles with varying height (say, slant-top type) can also be considered, with differential length of projection. As before, irregular-shaped obstacles can be approximated by means of some standard regular shape, either uniform or varying height, using the same model.

The total number of configuration space plots for the entire cluttered environment will depend on the number of slices each obstacle has and also thereby the average number of slices obtained. Since each of those sliced c-space will represent obstacle geometries, fully or partially, it is important to label the nodes of the obstacle-slices so produced. In general, if a particular obstacle has got 'k' slices, having 'n' nodes each, then a generalized node of that obstacle will be labeled as, 'n_k'. However, to avoid ambiguity, 'k' is considered alphabetic only. Figure 2 shows a representative obstacle with slices, where node '3_b' signifies the third node of the b[th.] slice, which is incidentally the second slice of the obstacle.

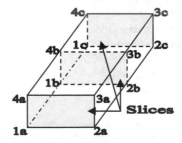

Fig. 2. Node numbering thematic for sliced obstacles

2.2 Paradigms of the C-space mapping algorithms in 2D *sliced* workspace

In the present work, *sliced* c-space maps have been generated considering a two degrees-of-freedom revolute type manipulator having finite dimensions (refer fig. 3) and an environment cluttered with polygonal obstacles. Both manipulator links and obstacles are represented as convex or concave polygons[4].

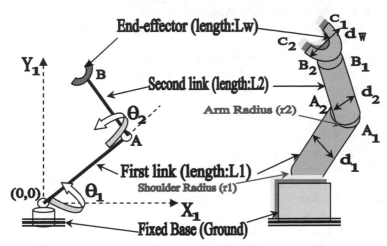

Fig. 3. Representative schematics of a two-link manipulator with revolute joints

The obstacles are considered to be *regular* in shape with fixed dimensions, having a well-defined shape (sectional view) in 2D, preferably convex, for easier calculations. This has been made purposefully as in most of the manufacturing and/or shop-floor activities *geometric* objects are being handled by the robot, for example, loading and unloading of components to/from the machine, handling of semi-finished components between machines, storage and retrieval of finished components into bins etc. The philosophy behind these mapping algorithms is to consider each complex obstacle as a boolean combination of various primitives, viz,. *'Point'*, *'Line'* and *'Circle'*. That is to say, if the obstacle is theoretically considered as a 'point', 'line' or 'circle' in shape in 2D, then colliding angle of the robot link(s) with those will be obtained and C-space maps can be drawn there from. These algorithms can also be applied for concave objects by considering the 'convex hull' of those and proceeding in the same manner taking that as the *new* obstacle. Similarly, irregular shaped objects can also be tackled with these models, in which *envelope* of the object is to be considered to get the nearest convex shape. Obviously accuracy of the results will suffer to some extent by this approximation, but it would be a reasonable solution for practical situations.

3. Details of the C-space mapping algorithms developed

3.1 Overview of the mapping algorithms

3.1.1 C-space transformation of "POINT"

This algorithm gives the C-space data for an obstacle, considered theoretically as a *point* in Cartesian space. The robot with line representation is being considered here. (refer fig.3).

---

[4] Concave obstacles are modeled as a combination of several convex polygons.

Algorithm:

1. Input robot base co-ordinates, link lengths, the specified *point*, range of joint-angles & angular resolution.
2. Calculate the distance of the *point* from robot base.
3. Initialize the loop with iteration i=1.
4. Check whether the first link is colliding.
5. Calculate the positional details of the first link (i.e. under colliding conditions), if step 4 is true.
6. Check the collision with the second link, if step 4 is false.
7. Calculate the colliding details of the second link, if step 6 is true.
8. Output the colliding angles for suitable cases.
9. i=i +1.
10. Continue till all combinations of joint - angles are checked.

3.1.2 C-space transformation of "LINE"
Here the obstacle has been considered as a regular polygon, bounded by several straight line-segments, as 'edges'. The algorithm is valid for obstacles having rectangle, square, triangle, trapezium etc. shaped cross-section in the vertical plane.

3.1.3 C-space transformation of "CIRCLE"
This algorithm for collision detection tackles the spherical obstacles with circular cross-section in the vertical plane. Various possible colliding conditions of the circular obstacle with the robot link(s) are depicted in fig. 4. It depends entirely on the location of the obstacle(s) with respect to the location of the robot in the workspace.

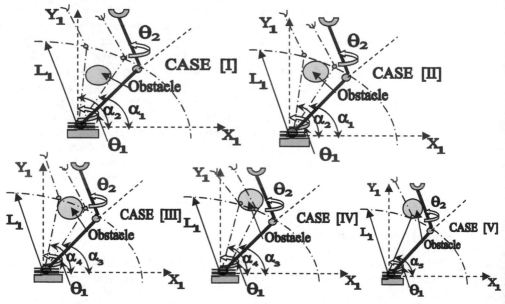

Fig. 4. Various possible colliding combinations of a two-link robot with circular obstacle

Figure 4 shows five different cases of collision, considering only the first link. However, the paradigm is valid for subsequent links also. These cases are:
Case I: Obstacle is fully within the range of the first link.
Case II: Obstacle is within the first link's range, touching the *range circle* internally.
Case III: Obstacle is collidable by the first link, with its centre inside the range of the first link.
Case IV: Same as before, but centre is outside the range of the first link.
Case V: Obstacle is touching the range circle of the first link externally.

3.1.4 C-space transformation with finite dimensions of the robot

Modifications of the previous algorithms for *POINT, LINE* and *CIRCLE* obstacles have been considered with finite dimensions of the robot arms (refer fig. 3). Also, planar 3-link manipulator is considered now, where the third link is nothing but the end-effector. Since planar movements are taken into account, only pitch motion of the wrist is considered along with shoulder and elbow rotations. The exhaustive list of input parameters for these cases will be as follows, viz. (i) Robot base co-ordinates (x_b, y_b); (ii) Length of the upper arm or shoulder (l_1); (iii) Length of the fore arm or elbow (l_2); (iv) Length of the end-effector or *wrist* (l_w); (v) Width of the upper arm (d_1); (vi) Width of the fore arm (d_2); (vii) Width of the end-effector (d_w); (viii) Radii of curvature for upper and fore arm $(r_1$ & r_2 respectively); (ix) The co-ordinates of the *Point*: (x_1, y_1) or *Line*: $[(x_1, y_1)$ & (x_2, y_2) as end - points] or *Circle*: [centre at (x_c, y_c) & radius : 'r'].

3.2 Analytical model of the mapping algorithms

The analytical models of various C-space mapping algorithms, described earlier, have been grouped into two categories, viz. (a) Model for *LINE* obstacles and (b) Model for *CIRCLE* obstacles. These are being described below.

3.2.1 Model for LINE obstacles

The ideation of collision detection phenomena for *Line* obstacle is based on the intersection of the line- segments in 2D considering only the kinematic chain of the manipulator. The positional information, i.e. co -ordinates (x_k, y_k) of the manipulator joints can be generalized as,

$$[x_k, y_k]^T = [x_b, y_b]^T + [\sum_{k=1}^{k=n} l_k \cos(\sum_{j=1}^{j=k} \theta_j), \sum_{k=1}^{k=n} l_k \sin(\sum_{j=1}^{j=k} \theta_j)]^T$$

$$\forall k = 1,2,3,\ldots\ldots\ldots,n \ \& \ \forall j = 1,2,\ldots\ldots, k \tag{1}$$

where,
$[x_b, y_b]^T$: Robot base co - ordinate vector in Newtonian frame of reference;
l_k : k^{th} link - length of the manipulator;
θ_j : j^{th} joint - angle of the manipulator.
The slope of the *line* (i.e. the edge of the obstacle) with (x_1, y_1) & (x_2, y_2) as end-points is given by,

$$m_0 = |(y_2 - y_1)| / (x_2 - x_1)| = |\Delta y / \Delta x| \tag{2}$$

Hence, the co - ordinates of the intersection point between the obstacle edge(s) and the robot link(s) are,

$$x_{int\_k} = (y_k - y_1 - m_k x_k + m_0 x_1) / (m_0 - m_k) \tag{3a}$$

$$y_{int\_k} = [m_0 m_k (x_k - x_1) - m_0 y_k + m_k y_1] / (m_k - m_0) \tag{3b}$$

where,

$[x_{int\_k} , y_{int\_k}]^T$: The intersection point vector for the k th. robot link with the *Line* and m_k : The slope of the k th. robot link.

For a 3-link planar revolute manipulator, we have, therefore,

$$m_1 = \tan \theta_{1i}, \text{where } \theta_{1i} \in [\theta_{1\_min} , \theta_{1\_max}]$$

and

$$m_2 = \tan (\theta_{1i} + \theta_{2j}), \text{ where } \theta_{2j} \in [\theta_{2\_min} , \theta_{2\_max}]$$

with θ_{1i} as defined earlier.

The above model can be extended likewise, considering finite dimension of the manipulator, i.e. having widths of the arms, viz. $\{ d_k \}$.

Nonetheless, considering finite dimensions, the generalized co-ordinates of the manipulator joints can be evaluated from,

$$[x_{Ak}, y_{Ak}]^T = [x_b, y_b]^T + [\sum_{k=1}^{k=n} l_k \cos (\sum_{j=1}^{j=k} \theta_j), \sum_{k=1}^{k=n} l_k \sin (\sum_{j=1}^{j=k} \theta_j)]^T$$

$$+ [d_k/2 \sin (\sum_{j=1}^{j=k} \theta_j) , - d_k/2 \cos (\sum_{j=1}^{j=k} \theta_j)]^T$$

$$\forall\ k = 1,2,3,\ldots\ldots\ldots,n\ \&\ \forall\ j = 1,2,\ldots\ldots,k \tag{4}$$

and,

$$[x_{Bk}, y_{Bk}]^T = [x_b, y_b]^T + [\sum_{k=1}^{k=n} l_k \cos (\sum_{j=1}^{j=k} \theta_j), \sum_{k=1}^{k=n} l_k \sin (\sum_{j=1}^{j=k} \theta_j)]^T$$

$$+[- d_k/2 \sin (\sum_{j=1}^{j=k} \theta_j) , d_k/2 \cos (\sum_{j=1}^{j=k} \theta_j)]^T$$

$$\forall k = 1,2,3,\ldots\ldots\ldots,n\ \&\ \forall\ j = 1,2,\ldots\ldots,k \tag{5}$$

Hence, the co-ordinates of the intersecting point(s) between the robot arm(s) and the *edges* of the obstacle(s) become,

$$x_{intA\_k} = (y_{Ak} - y_1 - m_k x_{Ak} + m_0 x_1) / (m_0 - m_k); \tag{6a}$$

$$y_{intA\_k} = [m_0 m_k (x_{Ak} - x_1) - m_0 y_{Ak} + m_k y_1] / (m_k - m_0); \tag{6b}$$

$$x_{intB\_k} = (y_{Bk} - y_1 - m_k x_{Bk} + m_0 x_1) / (m_0 - m_k); \tag{6c}$$

$$y_{intB\_k} = [\, m_0\, m_k\, (\, x_{Bk} - x_1\,) - m_0\, y_{Bk} + m_k\, y_1\,] / (\, m_k - m_0\,); \tag{6d}$$

considering two possible collisions per arm at the most.

3.2.2 Model for CIRCLE obstacles

With reference to fig. 4, let the following nomenclatures be defined,

$(\, x_b\,, y_b\,)$:   Robot base co-ordinates in Newtonian frame;

l_1 & l_2 :   Lengths of the first and second link of the robot respectively;

$(\, x_c\,, y_c\,)$:   Co-ordinates of the centre of the 'circle' obstacle and

r :   Radius of the 'circle'.

Let,

$$d = [\, (\, x_c - x_b\,)^2 + (\, y_c - y_b\,)^2\,]^{1/2} \tag{7}$$

If collision is detected with the first link of the robot, then the range of colliding angles for case I & II (i.e. α_1 & α_2) and for case III & IV (i.e. α_3 & α_4) will be evaluated as shown below,

$$\alpha_{2,1} = \tan^{-1}[(\, y_c - y_b\,) / (\, x_c - x_b\,)] \pm \tan^{-1}[\, r / (\, d^2 - r^2\,)^{1/2}\,] \tag{8}$$

$$\alpha_{4,3} = \tan^{-1}[(\, y_c - y_b\,) / (\, x_c - x_b\,)] \pm 2\tan^{-1}[\, (\, s - l_1\,)(\, s - d\,) / s(\, s - r\,)\,]^{1/2} \tag{9}$$

where,

$$2s = (\, l_1 + d + r\,) \tag{10}$$

Obviously, only one colliding angle, viz. α_5 will be obtained with case V, i.e.

$$\alpha_5 = \tan^{-1}[(\, y_c - y_b\,) / (\, x_c - x_b\,)] \tag{11}$$

When the collision occurs with the second link, i.e. all the values of the first joint-angle are collidable, the ranges for collision will be obtained as given below.

The co-ordinates of the start point of the second link, which serves as the *instantaneous base* of the robot, are given by,

$$[\, x_b'\,, y_b'\,]^T = [\, x_b\,, y_b\,]^T + [\, l_1 \cos\theta_1\,, l_1 \sin\theta_1\,]^T \tag{12}$$

The colliding range for the second link is between θ_{2s} and θ_{2f} against case I & II, where,

$$\theta_{2S} = \beta_1 - \theta_{1i}\, \text{and}\, \theta_{2f} = \beta_2 - \theta_{1i}\,, \forall\, \theta_{1i}\,, \text{where}\, \theta_{1\_min} \leq \theta_{1i} \leq \theta_{1\_max} \tag{13}$$

and,

$$\beta_{2,1} = \tan^{-1}[(\, y_c - y_b\,) / (\, x_c - x_b\,)] \pm \tan^{-1}(\, r / p\,) \tag{14}$$

where,

$$p = (\, h^2 - r^2\,)^{1/2} \tag{15}$$

and,

$$h = [\, (x_c - x_b')^2 + (\, y_c - y_b'\,)^2\,]^{1/2} \tag{16}$$

For case III & IV, the colliding range will be θ_{3S} to θ_{3f}, which can be evaluated as,

$$\theta_{3S} = \beta_3 - \theta_{1i} \text{ and } \theta_{3f} = \beta_4 - \theta_{1i}, \, \forall \, \theta_{1i}, \text{ where } \theta_{1i} \in [\, \theta_{1\_min}, \theta_{1\_max} \,] \tag{17}$$

and,

$$\beta_{4,3} = \tan^{-1}[(\, y_c - y_b')\, / \, (\, x_c - x_b')\,] \pm 2 \tan^{-1}[\, (\, s' - l_2)\, (\, s' - h)\, / \, s'\, (\, s' - r)\,]^{1/2} \tag{18}$$

where,

$$2\, s' = (\, l_2 + h + r) \tag{19}$$

For case V, the particular formidable value of the joint-angle is,

$$\theta_4 = \tan^{-1}[(\, y_c - y_b')\, / \, (\, x_c - x_b')\, - \theta_{1i} \tag{20}$$

The above equations need to be altered for collision checking with finite link dimensions of the manipulator, considering r1 and r_2 as the radii of curvature of the upper arm and fore arm respectively (refer fig. 3). The modifications required are: i] 'l_1' is to be replaced by l_u (= $l_1 + r_1$); ii] 'l_2' is to be replaced by l_f (= $l_2 + r_2$) and iii] 'r' is to be replaced by $r_{nj} = r + d_j\, /2$, $\forall$ j = 1,2, 3 corresponding to collision with upper arm, fore arm and the end-effector.

3.3 Illustrations using the developed algorithms

The c-space mapping algorithms have been tested with two different robot workspaces in 2D, as detailed below. Technical details of the robot under consideration for these workspaces are highlighted in Table 1.

Type of Robot Considered	Revolute
No. of Links	2
No. of Degrees -of- Freedom	2
Length of the Links	First Link: 5 units; Second Link: 4 units
Co-ordinates of the Robot Base	$x_b = 5$; $y_b = 5$
Ranges of Rotation of the Joints (Anticlockwise)	Case I: For Non-circular Polyhedral Obstacles Joint-angle 1: 0 to 360 deg. Joint-angle 2: 0 to 260 deg. Case II: For Circular Obstacles Joint-angle 1: -40 to 240 deg. Joint-angle 2: -180 to 180 deg.
Resolution of Joint Rotation	5 deg. (for both joints)

Table 1. Technical Features of the Robot Under Consideration

Figure 5 presents a 2D environment with non-circular polygonal obstacles with a revolute type robot located in-between.

After obtaining the relevant c-space data, various plots of joint-angle 1 vs. joint-angle 2 are made. It is to be noted that the final data-points are those, which are common values of formidable angles for the entire obstacle. Obviously, the points, which are inside or on the closed boundary of the curves, are 'collidable' combinations, and, hence, those should not be attempted for robot path planning.

Another case is studied, as shown in fig. 6, with circular obstacles only. There is a distinct difference in the appearance in the C-space maps between obstacles when collision occurs with the first as constrained with the same for the second link.

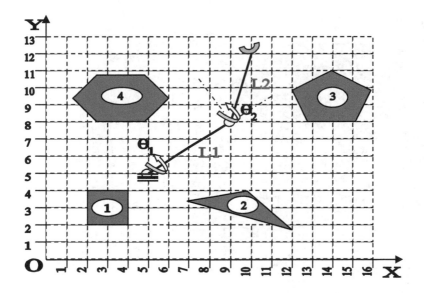

Fig. 5. A representative 2D environment congested with non-circular polygonal obstacles

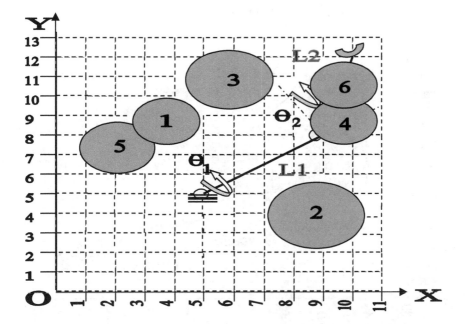

Fig. 6. A representative 2-D environment filled with circular obstacles

Figures 7 and 8 show the final c-space for the above two environments.

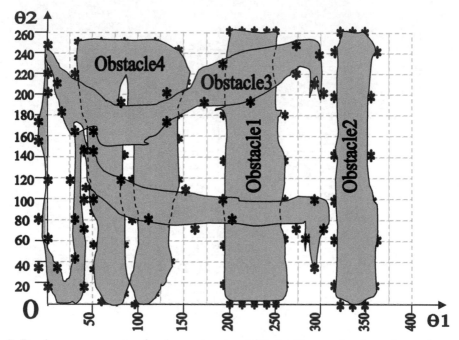

Fig. 7. Final c-space mapping for the environment filled with non-circular polygonal obstacles

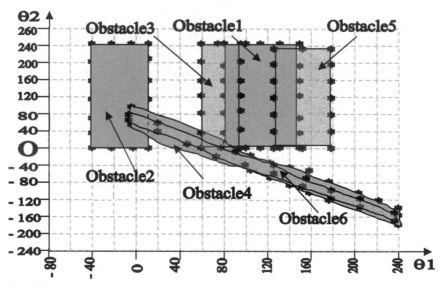

Fig. 8. Final c-space mapping for the environment congested with circular obstacles

3.4 C-space maps for higher dimensionality

The development of c-space is relatively simpler while we have two degrees-of-freedom robotic manipulators. As described in earlier sections, irrespective of the nature of the environment, we can generate simple planar maps, corresponding to the variations in the joint-angles (in case of revolute robots). Thus the mapping between task space and c-space is truly mathematical and involves computationable solutions for inverse kinematics routines. The procedure of generating c-space can be extrapolated to three degrees-of-freedom robots at most, wherein we get 3D plot, i.e. mathematically speaking, *c-space surface*. However, this procedure can't be applied to higher dimensional robots, having degrees-of-freedom more than three. In fact, c-space surface of dimensions greater than three is unrealizable, although it is quite common to have such robots in practice amidst cluttered environment. For example, let us take the case of a workspace for a seven degrees-of-freedom articulated robot, as depicted in fig. 9. Here we need to consider variations in each of the seven joint-angles, viz. θ_1, θ_2,......, θ_7 (the last two degrees-of-freedoms are attributed to the wrist rotations) towards avoiding collision with the obstacles.

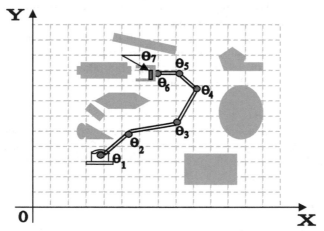

Fig. 9. A representative cluttered environment with seven degrees-of-freedom revolute robot

So, the question arises as how can we tackle this problem of realizing higher dimensionality of c-space mapping. Hence it is a clear case of building *composite c-space map*, with proper characterization of null space. We propose a model for this mapping in higher dimensionality as detailed below.

3.4.1 Lemma

1. Identify the degrees-of-freedom of the robot ['n'] and notify those.
2. The robot is conceptualized as a serial chain of micro-robots of two degrees-of-freedom each.
3. We will consider a total of 'k' planar c-space plots, where k = n/2 if 'n' is even and k = (n+1)/2 if 'n' is odd.
4. If 'n' is even, then the pairs for the planar c-space plots will be: $[q_1 - q_2]$, $[q_3 - q_4]$,......$[q_{n-1} - q_n]$, where 'q' denotes the generalized joint-variable of the manipulator.

5. If 'n' is odd, then the pairs for the planar c-space plots will be: $[q_1 - q_2]$, $[q_3 - q_4]$,......$[q_{n-2} - q_{n-1}]$, $[q_{n-1} - q_n]$.
6. In a way, we are considering several *virtual 2-link mini manipulators,* located at the respective joints of the original manipulator.
7. Out of the plots so generated, select the *most significant* c-space map.
8. One way of accessing the most significant c-space map is to consider finite measurement of the planar area of the c-space. A larger area automatically indicates more complex dynamics of the joint-variables so far as the collision avoidance is concerned.
9. Alternatively, significant c-space plots will be those having multiple disjointed loops, i.e. regions of formidable area. Individually the regions may be of smaller area, but the multiplicity of their occurrence adds complexity to the scenario.
10. Once the most significant c-space map is selected, the locations corresponding to 'S' and 'G' are to be affixed in that plot. This will be achieved using inverse kinematics routine from 'S'(x,y) and 'G' (x,y).
11. For the most significant plot so obtained, all joint-variables, except the two used in the plot will be *constant.* For example, if $[q_3 - q_4]$ plot is the most significant one, then q_1, q_2,......q_{n-1}, q_n are constant except q_3 and q_4.
12. In general, if $[q_i - q_j]$ plot be the most significant, then the set $\{q_1, q_2,......q_{n-1}, q_n\}$, except $[q_i - q_j]$, will be constant. And, the value of the set $\{q_1, q_2,......, q_{i-1}\}$ will be ascertained by the inverse kinematic solution of 'S' while the other set, $\{q_{j+1}, q_{j+2},.....,q_n\}$ will be determined by the inverse kinematic solution of 'G'.

3.4.2 Schematic of the model

Let us take an example of a seven degrees-of-freedom articulated robot, similar to one illustrated in fig. 9. According to the lemma proposed in 3.4.1, there will be four planar c-space plots, namely, $[\theta_1 - \theta_2]$, $[\theta_3 - \theta_4]$, $[\theta_5 - \theta_6]$ and $[\theta_6 - \theta_7]$. Figure 10 shows a sample view of these segmental c-space maps.

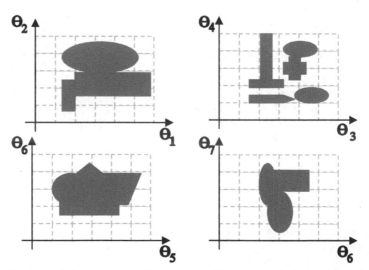

Fig. 10. Sample view of the composite C-space map for a seven degrees-of-freedom robot

As it may be apparent from fig. 10, either $[\theta_1 -- \theta_2]$ or $[\theta_3 -- \theta_4]$ plot can be significant, considering *larger area* or *presence of multiple loops* criterion (refer serial no. 8 & 9 of 3.4.1).

3.4.3 Generation of C-space plots: Concept of equivalent circle

In order to compute c-space data points for any particular combination of consecutive joint-variable pair for the higher dimensional robot, we would use a new concept, viz. the formulation of *Equivalent Circle* at the end of amidst the pair of links. Since we are considering *virtual two-link mini-manipulators* for the generation of c-space maps in pair, we would theoretically divide the links in two groups. The links, directly related to the generation of the specific c-space map, are termed as *active links*, while the others are known as *dummy links*. The philosophy of this equivalent circle is to re-represent the higher dimensional manipulator with only the active links and the joints therein, interfaced with circular zone(s) either at the bottom of the first active link or at the tip of the second active link. In general, the equivalent circles are constructed considering full rotational freedom of all the dummy links, located before / after the active links.

For example, if we wish to generate $[\theta_1 -- \theta_2]$ plot for the seven d.o.f. manipulator, then the *equivalent circle* alias *equivalent formidable zone* is to be constructed adjacent to the end the second link and circumscribing the remaining links. Figure 11 schematically presents the concept of *equivalent formidable zone*, with first two links as active links for a seven d.o.f. manipulator.

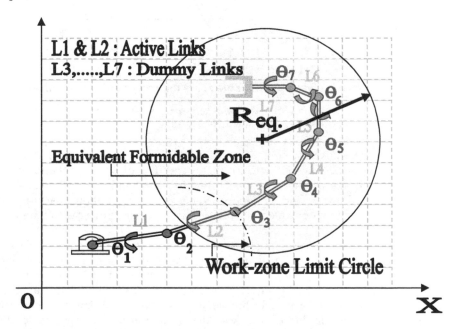

Fig. 11. Schematic view of equivalent formidable zone for a seven d.o.f. manipulator

However it is possible to have two *equivalent formidable zones* in cases where some intermediate links are considered for c-space plots. For example, if the third & fourth links of the manipulator become active links, then there will be two formidable zones, as illustrated in fig. 12. Of course, as per this proposition, there can't be more than two formidable zones for any higher dimensional manipulator.

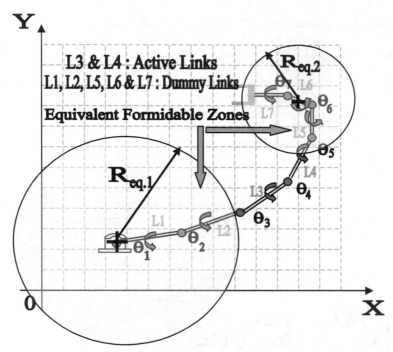

Fig. 12. Occurrence of two equivalent formidable zones for a seven d.o.f. manipulator

It is essential to locate the center of the *equivalent circle* vis-à-vis its radius (equivalent radius, $R_{eq.}$), in order to start computing for colliding combinations. However, the formulations for equivalent radii are not same in the two cases, as cited in figs. 11 & 12. In fact, the center of the equivalent circle, for cases wherein active links are followed by dummy links, will be at the base of the manipulator and its radius will be the summation of the lengths of the dummy links till we reach the first active link. For example, $R_{eq.\,1}$ in fig. 12 will be the added sum of L1 & L2. Figure 13 shows the computational backup for the evaluation of the *radius* of the equivalent formidable zone /circle in this case, nomenclated as *equivalent radius [type I]*.

Although finding center and calculating equivalent radius [type I] is straight -forward, evaluation of the new *base* for the *virtual two-link robot* is critical. For example, as shown in fig. 12, we need to find out the possible location of the base for the virtual robot, comprising of L3 & L4. This has been explained in fig. 14, wherein we adopt a methodology, called *Least Path*. The least path(s) is/are the straight line-segment(s) joining the centers of equivalent circle [type I] and the obstacles in the vicinity of the virtual robot. The respective locations for the base of the virtual robot will be the point of intersection of the least path and the equivalent circle. Thus, as per fig. 14, there will be three base-points, viz. 'C_p'', 'C_q'' & 'C_r'' corresponding to three obstacles, vide 'p', 'q' & 'r', which are situated within the limit-zone of the virtual robot with links L_k & L_{k+1}. However, this location of the base can range between two extreme points, as detailed in the inset of fig. 14. Two cases may appear here,

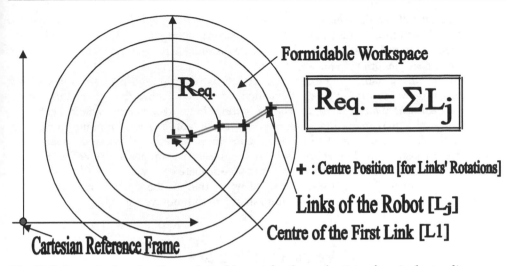

Fig. 13. Schematic showing the analytical layout for the evaluation of equivalent radius [type I]

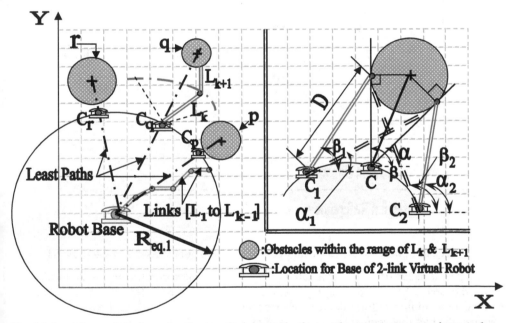

Fig. 14. Locations of the base for the two link virtual robot with respect to equivalent circle [type I]

viz. when the obstacle is a] within the range of both L_k & L_{k+1} and b] outside the range of L_k. Now, considering the first link of the virtual robot is just able to touch the obstacle we can get two extreme positions for the base, e.g. 'C_1' & 'C_2'. As explained earlier, the point 'C' is the other location for the base, situated on the least path. From geometry, we can assign 'D', which is numerically equal to either L_k or L_{k+1}, depending upon which link is colliding with that obstacle. The collidable range of joint-angles, corresponding to 'C', 'C_1' & 'C_2' will be (β-α), (β_1-α_1) & (β_2-α_2) respectively. Thus, in general a formidable range from α_2 to β_1 should be selected for c-space mapping in (θ_k --θ_{k+1}) plot.

In contrary to this situation, the other one, namely where dummy links are followed by active links, is more intricate so far the thematic is concerned. Figure 15 explains this case of evaluating *equivalent radius [type II]*, the corresponding circular zone being *optimized* between two extremes, viz. maximum and minimum formidable zones. The minimum formidable zone is a circular space, tangent to the work-zone limit circle at 'A' while the maximum formidable zone is a semi-circular area, with two opposite extremities as 'Z_1' & 'Z_2'. Several feasible formidable zones are theoretically possible in-between, with pair of *chordal end-points* as [B_1 – B_1^*] or [B_2 – B_2^*] etc. It may be noted that all the three formidable zones, namely the minimum, maximum & optimum, share the common vertex 'V'.

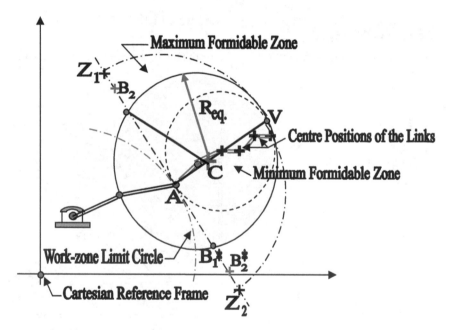

Fig. 15. Schematic showing the disposition of the equivalent formidable circle [type II]

The *equivalent radius [type II]* is evaluated using geometrical attributes, as detailed in fig. 16. Here, the points 'A', 'C' & 'C_m' represent the locations of the centers of the maximum, equivalent & minimum formidable zones respectively. As evident from the figure, the ratio between the two line-segments, viz. the semi-chordal length of the equivalent circle and the radius of the maximum formidable zone is 'k', where 0<k<1. In-line with the numerical evaluation of '$R_{eq.}$', the location of the center, 'C' can be determined also.[5]

---

[5] The location of the center is determined by evaluating the length of the line-segment, AC, which is numerically equal to $[(1-k^2)/2]\Sigma L_j$ and it is also at a distance of $(k^2/2)\Sigma L_j$ from the center of the minimum formidable zone, i.e. 'C_m'.

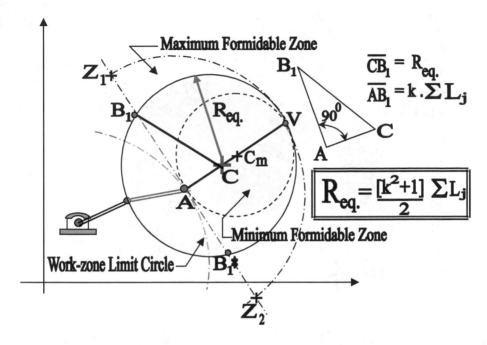

Fig. 16. Schematic showing the analytical layout for the evaluation of equivalent radius [type II]

3.5 Evaluation of C-space plots for higher dimensional robot: Example

Here we will study a specific case for a seven degrees-of-freedom robot, amidst a 2D cluttered environment as shown in fig. 17. The technical parameters of the robot, comprising link-lengths, $\{l_i, \forall i = 1,2,...7\}$ and joint limits $\{\theta_i, \forall i = 1,2,...7\}$, are highlighted in Table 2. Circular obstacles are considered for simplicity in computations. The locations of the respective centers and diameters of the obstacles (expressed in suitable units) are presented in Table 3.

Type of Robot Considered	Revolute
No. of Links	7
No. of Degrees -of- Freedom	7
Length of the Links [in suitable units]	l_1=15.52; l_2=14.87; l_3=13.04; l_4=12.54; l_5=9.0; l_6=5.22 & l_7=8.0
Co-ordinates of the Robot Base	x_b = 10; y_b = 10
Ranges of Rotation of the Joints (Anticlockwise)	θ_1: -20⁰ to 140⁰; θ_2: 0⁰ to 120⁰; θ_3: 0⁰ to 80⁰; θ_4: -5⁰ to 90⁰; θ_5: 10⁰ to 50⁰; θ_6: -5⁰ to 55⁰; θ_7: 5⁰ to 35⁰
Resolution of Joint Rotation	3 deg. (for all joints)

Table 2. Technical Facets of the Higher Dimensional Robot Under Consideration

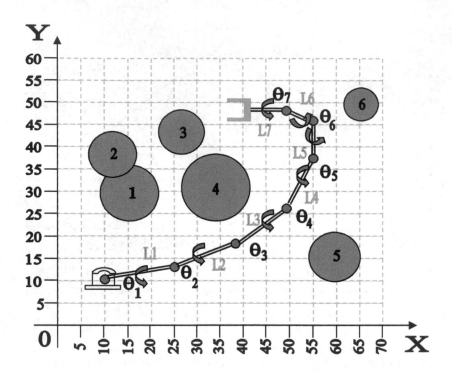

Fig. 17. Workspace layout of the seven degrees-of-freedom articulated robot

Obstacle No.	Diameter	Location of Centre
1	15.88	(x =15, y = 30)
2	13	(x = 12.5, y = 37.5)
3	12.7	(x = 27.5, y = 42.5)
4	19	(x = 35, y = 30)
5	14	(x = 60, y = 15)
6	9.53	(x = 65, y = 50)

Table 3. Obstacle Signature, as per Workspace Layout of Figure 16

Now, in this case of seven degrees-of-freedom revolute robot, we will have four different c-space plots, namely, $[\theta_1 -- \theta_2]$, $[\theta_3 -- \theta_4]$, $[\theta_5 -- \theta_6]$ & $[\theta_6 -- \theta_7]$. All of these plots use collision-detection algorithms, described earlier, for each of the obstacles separately, taking into account the concepts of *equivalent circles*. These c-space plots are illustrated in figure 18. A gross estimate reveal that $[\theta_1 -- \theta_2]$, $[\theta_3 -- \theta_4]$, $[\theta_5 -- \theta_6]$ & $[\theta_6 -- \theta_7]$ plots occupy a planar area measuring (160x120), (80x95), (40,60) & (60x30) sq. units respectively. It is evident from fig. 18 that although complexity-wise both $[\theta_1 -- \theta_2]$ and $[\theta_3 -- \theta_4]$ plots are roughly at par, but the former is to be selected as the most significant c-space plot as it is also the largest in size.

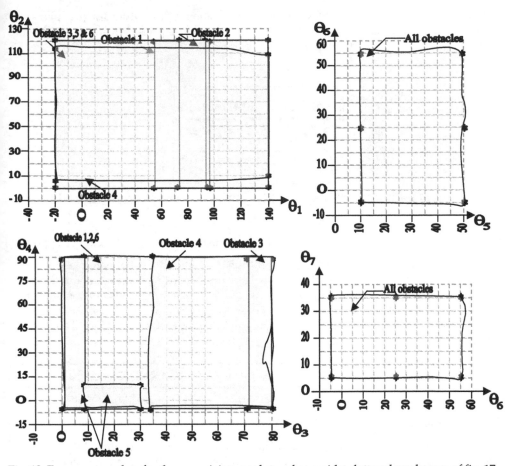

Fig. 18. Four c-space plots for the seven joint revolute robot amidst cluttered workspace of fig. 17

It may be noted that while c-space plot for a 2 d.o.f. robot (working in 2D or 3D task-space) can be composed of irregular non-geometric shapes (refer fig. 7), the same for higher d.o.f. robots are perfectly geometric (refer fig. 18). This is happening because of the incorporation of the concept of 'formidable zones' for higher dimensional robots, wherein we are deliberately allowing the collidable zone to engulf more regions in the c-space. In fact, in most of cases for higher d.o.f. robots, the c-space zones are perfect rectangular in shapes, between the minimum & maximum limits of the participating joint-angles. For example, in fig. 18, (θ_1 vs. θ_2) c-space slice plot the final rectangle is constituted between 4 vertices, viz. θ_{1_min}, θ_{1_max}, θ_{2_min} & θ_{2_max}.

4. Safe path in configuration space: Logistics & algorithm

4.1 Perspective
Based on the formation of c-space maps, collision-free paths are to be ascertained in 2D as well as in 3D. We will analyze the gamut in four functional *quadrants*, which have been

conceptualized from the point of view of a] disposition of the obstacles (i.e. planar or spatial) and b] kinematics of the manipulator (i.e. its degrees-of-freedom). Following checker-box illustrates the situation pictorially (refer fig. 19).

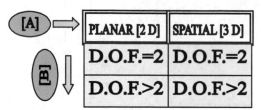

[A] : Disposition of Obstacle   [B] : Degrees-of-freedom of the Robot

Fig. 19. Schematic of the robotic path planning scenario using Configuration Space approach

In fact, the methodologies to be adopted for different situations of robot path planning vary significantly and those depend on the task-space nature (i.e. planar or spatial) and the robot kinematics (i.e degrees-of-freedom). Besides, the 3D path planning is also dependent upon the nature of the slices produced from the task-space. Those can be identical in nature or non-identical. Table 4 presents the scenario, highlighting the methods used for the collision-free path planning,

Parameter	COLLISION-FREE PATH PLANNING					
Task-space	2 D [Planar]		3 D [Spatial]			
Robot Type	D.O.F. = 2	D.O.F. > 2	D.O.F. = 2		D.O.F. > 2	
Method Used for Path Planning	a] C-space Map & b] V-graph	a] (Multiple) C-space Maps ⇒ b] Most Significant C-space Map & c] V-graph	**Identical Slice**	**Non-identical Slice**	**Identical Slice**	**Non-identical Slice**
			a] Sliced C-space Maps ⇒ b] V-graph based Paths & c] Final Path	a] Sliced C-space Maps ⇒ b] V-graph based Paths &c] *Intersection* of the Feasible Paths	a] (Multiple) C-space Maps ⇒b] Most Significant C-space Map & c] V-graph for each slice & d] Final Path	a] (Multiple) C-space Maps ⇒b] Most Significant C-space Map & c] V-graph for each slice & d] *Union* of the Feasible Paths

Table 4. Illustrative Summary of the Methods Used for Collision-free Path Planning of Robots

4.2 Logistics of visibility graph formulation: Our model

The workspace of the robot has been modeled by formulating the *Visibility Matrix* of the *visibility graph*[6] of the cluttered environment. This matrix has been conceived as a kind of 'adjacency' relationship and framed by knowing the necessary visibility information about the nodes of the graph. Figure 21 illustrates the visibility matrix, [Vij], as developed from the environment shown in fig. 20, which depicts a typical sub-visibility graph, generated out of several polygonal obstacles known a-priori (i.e. with pre-fixed locations).

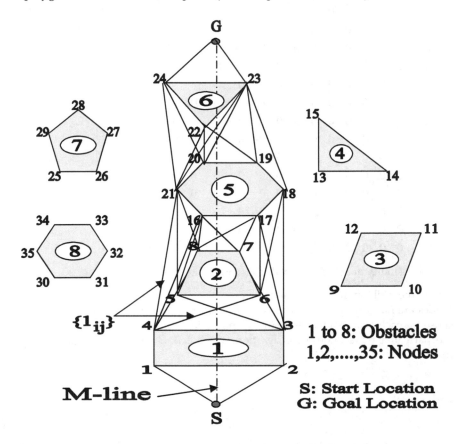

Fig. 20. A representative sub-visibility graph in two dimensions

The matrix is of the order (N + 2) x (N + 2), 'N' being the total number of intermediate nodes of the graph. The full matrix is formed by adding the 'start' (S) and 'goal' (G) nodes, making [V ij] a square matrix. Each row of [Vij] gives the visibility information of that very node. For example, the fifth row of the matrix signifies that the node no. 4 (considering 'S' as the first node) can 'see' nodes 5, 6, 8, 16 & 21 only.

---

[6] In order to reduce computational burden, 'Sub-visibility Graph' technique has been adopted in the present study. It considers only those obstacles in the robot workspace, which collide directly with the M-line.

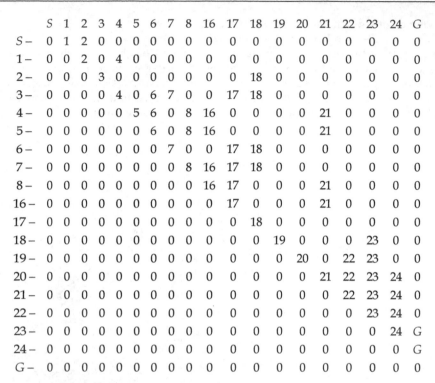

	S	1	2	3	4	5	6	7	8	16	17	18	19	20	21	22	23	24	G
S –	0	1	2	0	0	0	0	0	0	0	0	0	0	0	0	0	0	0	0
1 –	0	0	2	0	4	0	0	0	0	0	0	0	0	0	0	0	0	0	0
2 –	0	0	0	3	0	0	0	0	0	0	0	18	0	0	0	0	0	0	0
3 –	0	0	0	0	4	0	6	7	0	0	17	18	0	0	0	0	0	0	0
4 –	0	0	0	0	0	5	6	0	8	16	0	0	0	0	21	0	0	0	0
5 –	0	0	0	0	0	0	6	0	8	16	0	0	0	0	21	0	0	0	0
6 –	0	0	0	0	0	0	0	7	0	0	17	18	0	0	0	0	0	0	0
7 –	0	0	0	0	0	0	0	0	8	16	17	18	0	0	0	0	0	0	0
8 –	0	0	0	0	0	0	0	0	0	16	17	0	0	0	21	0	0	0	0
16 –	0	0	0	0	0	0	0	0	0	0	17	0	0	0	21	0	0	0	0
17 –	0	0	0	0	0	0	0	0	0	0	0	18	0	0	0	0	0	0	0
18 –	0	0	0	0	0	0	0	0	0	0	0	0	19	0	0	0	23	0	0
19 –	0	0	0	0	0	0	0	0	0	0	0	0	0	20	0	22	23	0	0
20 –	0	0	0	0	0	0	0	0	0	0	0	0	0	0	21	22	23	24	0
21 –	0	0	0	0	0	0	0	0	0	0	0	0	0	0	0	22	23	24	0
22 –	0	0	0	0	0	0	0	0	0	0	0	0	0	0	0	0	23	24	0
23 –	0	0	0	0	0	0	0	0	0	0	0	0	0	0	0	0	0	24	G
24 –	0	0	0	0	0	0	0	0	0	0	0	0	0	0	0	0	0	0	G
G –	0	0	0	0	0	0	0	0	0	0	0	0	0	0	0	0	0	0	0

Fig. 21. The visibility matrix developed from the environment shown in fig. 20

Features of the visibility matrix are listed below:

i.   Nodes are numbered in ascending order from 'S' to 'G' maintaining counter-clockwise sense for each obstacle. Backtracking of the nodes is not allowed. Symbolically, if 'n_r' is the particular node under consideration which can see nodes $n_{j1}, n_{j2},$............, $n_{jp},$...., n_{jk} then:

$$n_{j1} > n_r$$

and

$$n_{r-1} \le n_{jp} \le N \tag{21}$$

where 'n_{jp}' is any general node of the graph, *visible* by the node 'n_r' and 'N' is the goal node of the nodal series, namely $1,2,3,$.........,N.

ii.   Numbering of nodes is exhaustive and independent of the obstacle nomenclature. In other words, 'n_{jp}' is invariant to obstacle number and also does not have any correlation with any specific obstacle geometry in any form.

iii.   The entire matrix has got two parts symmetrically disposed off by the diagonal. Because of the reason stated earlier, the generalized element, 'a_{ij}' of $[Vij]$, which signifies the visibility information of the $j^{th.}$ node as *viewable* with the $i^{th.}$ node as *viewer*, and can be expressed mathematically as:

$$a_{ij} = 0 \text{ or } j \,, \forall j \ge i$$

$$a_{ij} = 0 \ \forall \ j \le i$$

$$a_{i1} = 0 \ \forall \ i$$

$$a_{Nj} = 0 \ \forall \ j \tag{22}$$

iv. For simplicity in computation, the original square matrix [Vij] of order (N+2) has been truncated to a square matrix of order (N+1), by deleting the first column and the last row. Since all the elements of these row and column are zero as per proposition, this modification will not alter the final result. Obviously, structure of this reduced matrix depends on the modeling of the robot environment and the total number of nodes present.

v. It has been found from the composite evaluations that the total number of computations required for the path planning algorithms is slightly less than O (N²), 'N' being the total number of nodes in the visibility graph, corresponding to the workspace under investigation.

vi. Essentially the visibility matrix reduces to the following structure in its most practical form, viz.:

$$[Vij] = \begin{bmatrix} 0 & 0 & d & d & d & 0 \\ 0 & 0 & 0 & 0 & d & d \\ 0 & 0 & 0 & 0 & d & 0 \\ 0 & 0 & 0 & 0 & 0 & d \\ 0 & 0 & 0 & 0 & 0 & d \\ 0 & 0 & 0 & 0 & 0 & 0 \end{bmatrix}$$

where, 'd' signifies the locations for non-zero entry. The left triangular matrix will be zero in a majority of cases, alongwith the diagonal elements. However, the visibility matrix may include some non-zero entries too corresponding to 'sculptured obstacles' with multiple vertices. With this structure of [Vij], which is time-optimized, the nodal lines can be computed as:

$$\{Lij\} = \sum_{i=1}^{i=N} C_i \left(a_{ij} \ ^* \right) \tag{23}$$

where, {Lij} is the nodal line vector, $(a_{ij} \ ^*)$ is the non-zero $[a_{ij}]$ and $C_i \ (a_{ij} \ ^*)$ is the cardinality of $a_{ij} \ ^*$. For example, $C_i \ (a_{ij} \ ^*)$ for the 'S' node (i.e. the first row of [Vij] with i = 1) becomes 2.

In addition, the total number of imaginary lines in the visibility graph can be computed as:

$$\{Lij^*\} = \{Lij\} - \sum_{k=1}^{k=N} d_k \tag{24}$$

where, 'd_k' is the number of edges of the $k^{th.}$ obstacle.

vii. A quick estimation of the computational time for the algorithmic path planning using a finalized visibility graph of the robot workspace reveals the following:

$$\tau \propto \{Lij\} \text{ or } \{Lij^*\} \tag{25}$$

where, $'\tau'$ is a factor representing the computational time and the memory requirement factor, say ξ, will be as follows:

$$\xi \propto \{Lij^*\} \tag{26}$$

Also the range of $'\tau'$ may be estimated for all practical computational situations as,

$$O(N) < \tau \le O(N^{3/2}) \tag{27}$$

It is to be noted that the lower bound of $'\tau'$ is certainly beyond $O(N)$, while the upper bound can marginally reach $O(N^{3/2})$, as 'N' tends to some larger value.

Regarding circular obstacles in 2D plane, a trade-off has been attained between the approximation to the nearest polygonal shape and the computational burden. For example, a perfectly circular-shaped object can be approximated with a circumscribing square-shaped object, but it will be about 80 % less efficient in comparison to an approximation with an octagonal shape. Hence, the decision for the optimal selection for approximation for a circular obstacle remains with the overall complexity of the visibility graph, i.e. essentially the value of 'N' generated thereby.

4.3 Development of the path planning algorithm in 2D plane

The new heuristic algorithm, namely, Angular Deviation Algorithm, has been developed to obtain near-optimal path for the manipulator amidst obstacles in a planar workspace. The formulation of the heuristics and subsequently the solution phase are based on A* search technique, in general. Following legends have been used in formulating the algorithm.

S: The 'start' location of the robot end-effector (in Cartesian or C-Space);

G: The 'goal' location of the robot end-effector (in Cartesian or C-Space);

SG : The imaginary line joining 'S' and 'G';

L_{M-Line}: The geometric length of the imaginary line joining 'S' & 'G', i.e. the 'M-Line';

x : An intermediate level in the graph search process;

V_{x_j} : j^{th} visible node from x^{th} node in the visibility graph;

$x\ V_{x\_j}$: The nodal line, joining the xth node and the jth visible node (from x^{th} node);

i : Iteration number of the graph search process.

This algorithm relies on *angular deviation* as the necessary computing heuristic, which is in-built in nature. It considers each line-segment, $\{l_{ij}\}$, where, $l_{ij} \in \{1\}$, joining the nodes, say, n_i and n_j, where $(n_i, n_j) \in \{n\}$, $\forall\ i, j \in I$, of the sub-visibility graph and computes the angular deviation of that $\{l_{ij}\}$ with respect to the M-line. The logic of this algorithm is to minimize the Angular Heuristic Function, as generated from the angular deviations, at each level of searching. Hence, in a cluttered environment the near-optimal path can be chosen by considering only those line-segments, which are comparatively closer to the M-line.

Steps:

1.   Initialize the search with i=1.
2.   For i=1, loop starts with 'S': note $V_{S\_1}, V_{S\_2}, \dots\dots\dots, V_{S\_p}$.
3.   Compute: $\alpha_{S\_1} = Ang(SG, SV_{S\_1}); \alpha_{S\_2} = Ang(SG, SV_{S\_2}),\dots\dots\dots \alpha_{S\_p} = Ang(SG, SV_{S\_p})$.
4.   Check: min. $(\alpha_{S\_1}, \alpha_{S\_2},\dots\dots\dots, \alpha_{S\_p})$.
5.   If $V_{S\_p} = 'G'$, then stop.

Else,

6. $i = i + 1$.
7.   Begin the next level of search from the $x^{th.}$ node with min.(α_{S_x}).
8.   Compute: { α_{x_j} } = { Ang (SG, XV_{x_j}) }, where $x < j \leq N$, 'N' being the total number of nodes in the graph.
9.   Go on searching likewise till 'G' occurs and finally note the nodes of the optimal path, so achieved.

4.4 Paradigms of path planning in 3D space for two degrees-of-freedom robots
4.4.1 Model for C-space slices and path generation

The concept of discretization of robot workspace in preferential 2D slices has been applied for generating safe path in a spatial manifold. As a result, multiple c-space slices are generated depending upon the features of the individual slices. Safe paths are then determined, using the *Angular deviation algorithm*, separately for each such c-space slice produced. Thus, if a spatial workspace is segregated in 'k' slices, then there will be 'k' c-space slices also. Figure 22 schematically presents the view of the *sliced c-space maps* for a known environment.

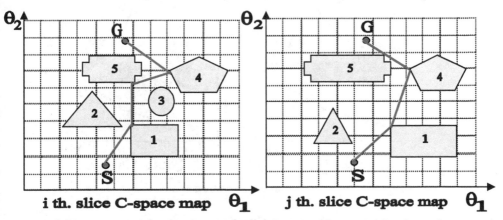

Fig. 22. Schematic view of the sliced c-space maps for a two degrees-of-freedom robot workspace

It may be noted that we need to generate *c-space slices*, as shown in fig. 22, for all the slices equally since our c-space mapping algorithms are in 2D and those are based on planar collision avoidance principles concerning *Point, Line* & *Circle* obstacles. Since the total number of slices for a specific environment is fixed a-priori, it may so happen that a particular slice of an object is not falling under the obstruction zone of the robot. In that case, that particular slice of that object will not appear in the corresponding c-space slice (e.g. refer $j^{th.}$ slice c-space map of fig. 22, wherein the slice for obstacle 3 is absent). Nonetheless, the axiom followed is *if the last slice of any obstacle is collidable, then all its previous slices ought to be. But the vice-versa is not true.*

Once the requisite c-space slice maps are generated, we need to evaluate the *safe* paths in each of these sliced c-space maps using the visibility-based Angular deviation algorithm. Thus a set of paths will be obtained and the cardinality of the set will be equal to the number of slices. Finally we will take the *intersection* of the possible paths, as only intersection set will represent the optimal path in true sense. However, intersection won't be the solution for situations where slice(s) for obstacle(s) is/are absent in a particular c-space slice. For example, with reference to fig. 22, considering two c-space slices in total, we have to follow the path shown in the first map, i.e. the ith slice map.

On the other hand, in situations where all the object-slices are present in all the c-space slices, then intersection can be advantageous, as we can omit longer routes via nodes in certain instances. A typical case is exemplified in fig. 23. Here a particular object is shown to have slightly different geometries and the path between 'S' & 'G' also varies accordingly as shown. In case (a), we are unable to consider 2' as 'node', because of the proposition of visibility graph and the path (viz. S$\Rightarrow$ ......1$\Rightarrow$2$\Rightarrow$3$\Rightarrow$G) is bound to pass through the stipulated node 2 only. However, node 2' lies very much inside the boundary of the c-space obstacle and in fact, it is closer to 'G' as compared to node 2. Thus the path shown in case (b) is the optimal solution (viz. S$\Rightarrow$ ....1$\Rightarrow$2'$\Rightarrow$G) and in fact, it is also the intersection of the two alternatives.

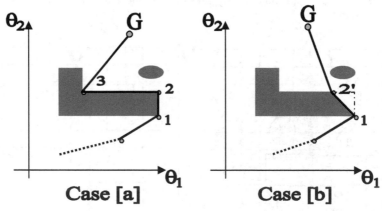

Fig. 23. Selection of path using intersection of alternatives (paths)

4.4.2 Quantitative evaluation of C-space slice points

In order to evaluate the c-space points mathematically, the relevant algorithm generates the intercept co-ordinates (X & Y) corresponding to each 'slice', which are governed by the rotational range of the robot waist and the finite resolution of the waist rotation. The program is applicable only to rectangular 3D solid obstacles, e.g. cube, rectangular parallelopiped, pyramid etc., either directly or after duly transformed from spherical or semi-spherical obstacles. The model is being illustrated schematically through fig. 24.

The relevant formulation vis-à-vis algorithm of the concerned model is described in detail below. Consider figure 24, let:

w : Width of the obstacle;

d : Distance between the robot and the obstacle ;

(x_b , y_b) : Co-ordinates of the robot base;

(x_1 , y_1) & (x_2 , y_2) : Co-ordinates of the edge of the obstacle in consideration in 2D elevation;

$\theta_{b\_max.}$ & $\theta_{b\_min.}$: Maximum and minimum values of the base rotational edge ;

θ_S : Slicing value of the base rotational angle.

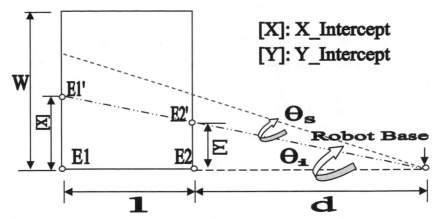

Fig. 24. Schematic representation of the dimensional metrics of a 2D 'Slice'

The range of base rotation is computed as,

$$r_{base} = (\theta_{b\_max.} - \theta_{b\_min.}) \tag{28}$$

Length of the 'edge' of the obstacle is,

$$l = [\, (x_2 - x_1)^2 + (y_2 - y_1)^2\,]^{1/2} \tag{29}$$

Now, only one slice is possible, if,

$$d\,(\pi\,\theta_S\,/\,180\,) > w \tag{30}$$

In situations where more than one slice is possible for a particular obstacle, two cases can appear.

Case I: Robot base is in-line with the obstacle

In this case, let $\quad S = d\,(\pi\,\theta_i\,/\,180)$, where $\theta_S \le \theta_i \le r_{base}$.
If $S \le w$, then 'slice' is possible and co-ordinates of the intercept of the slices are given by,

$$x_{int\_1} = x_1 + [\, l \tan \theta_i\,/\,(\,l/d)\,]\,[\,1 + (\,l/d\,)] \tag{31}$$

$$y_{int\_1} = y_1 + [l \tan \theta_i\,/\,(\,l/d)\,]\,[\,1 + (\,l/d\,)] \tag{32}$$

$$x_{int\_2} = x_2 + [l \tan \theta_i\,/\,(\,l/d)\,] \tag{33}$$

$$y_{int\_2} = y_2 + [l \tan \theta_i\,/\,(\,l/d)\,] \tag{34}$$

where, (x_{int_1}, y_{int_1}) and (x_{int_2}, y_{int_2}) are one set of co-ordinates corresponding to one slice. With θ_i varying within its range with an increment of θ_S, the other set of slice co-ordinates will be obtained.

Case II: Robot base is not in-line with the obstacle

In this case,

$$\theta_{in} = \tan^{-1}[\,|\,(y_2 - y_b)\,/\,(x_2 - x_b)\,|\,] \tag{35}$$

Also,

$$S' = d \left[\pi(\theta_j - \theta_{in}) / 180 \right], \forall j, \text{ where } \theta_{in} \leq \theta_j \leq r_{base}.$$ (36)

If $S' \neq 0$ and $S' \leq w$, then slice is possible and co-ordinates of the intercept of the slices are given by,

$$x_{int\_1n} = x_1 + \left[1 \tan (\theta_j - \theta_{in}) / (1/d) \right] \left[1 + (1/d) \right]$$ (37)

$$y_{int\_1n} = y_1 + \left[1 \tan (\theta_j - \theta_{in}) / (1/d) \right] \left[1 + (1/d) \right]$$ (38)

$$x_{int\_2n} = x_2 + \left[1 \tan (\theta_j - \theta_{in}) / (1/d) \right]$$ (39)

$$y_{int\_2n} = y_2 + \left[1 \tan (\theta_j - \theta_{in}) / (1/d) \right]$$ (40)

As before, (x_{int_1n}, y_{int_1n}) and (x_{int_2n}, y_{int_2n}) are one set of co-ordinates corresponding to one slice. With θ_j varying within its range with an increment of θ_S, the other set of slice co-ordinates will be obtained. Selection of 'safe' nodes of the robot end-effector in the 3D space depends on the elemental results, as obtained from the corresponding planar analysis. Symbolically, if the collision-free path for one particular 'slice' is represented as,

$$\{ P_{S\_k} \} = \{ S, X_1, X_2, \ldots\ldots\ldots\ldots, G \}_{S\_k}$$ (41)

where, $X_1, X_2, \ldots\ldots\ldots\ldots$, represent the serial number of the 'safe' nodes (may not be in the same order as the node nos., viz, 1,2,3,........) and 's_k' is the kth slice, then the final path will be the union of all such feasible combinations, viz.,

$$\{ P \} = \bigcup_{k=1}^{k=n} \{ P_{S\_k} \}$$ (42)

where, 'n' is the total number of slices generated.

4.5 Evaluation of path in 3D for higher dimensional robot
As depicted in fig. 19, evaluation of the collision-free path in 3D will depend on the degrees-of-freedom of the robot (i.e. whether d.o.f. =2 or >2). Nonetheless, in both the cases we need to fragment the 3D task-space in multiple slices, which may or may not be identical to one another. Thus, we will arrive at a situation wherein we have to deploy different strategies to evaluate a safe path. These models are described below.

4.5.1 Model for obtaining safe path for identical slices
The first and foremost pre-requisite of evaluating a collision-free path in this case is to have the *most significant 2D c-space slice map* for the robotic workspace under consideration, as described in 3.4.2. Once the critical c-space slice map is earmarked, the next task is to pin-point the corresponding locations for 'S' & 'G' in that map. This can be achieved by using the inverse kinematic solution for 'S' & 'G' and subsequent mapping from task space to joint space. For example, consider again the case of a seven degrees-of-freedom robot shown in fig. 9, wherein say the $[\theta_3 -- \theta_4]$ map is the most significant. Thus, the corresponding visibility graph of the environment will have non-identical 'S' & 'G' signatured as, S: $(\theta_3 = \alpha^0, \theta_4 = \beta^0)$ and G: $(\theta_3 = \gamma^0, \theta_4 = \delta^0)$. We will assume that the set of other joint-angles, i.e. $\{\theta_1, \theta_2, \theta_5,$

θ_6, θ_7} to be constant throughout the process of path generation. Now, by using the developed path planning algorithm, we will finally get a collision-free optimal path between S: (α,β) and G: (γ,δ). The generalized representation of co-ordinates of any two nodes (say N_i & N_j) of the said path will be as follows,

$$N_i \equiv \{(\theta_1 = c_1, \theta_2 = c_2), \theta_3 = x_i, \theta_4 = y_i, (\theta_5 = c_5, \theta_6 = c_6, \theta_7 = c_7)\}$$

and

$$N_j \equiv \{(\theta_1 = c_1, \theta_2 = c_2), \theta_3 = x_j, \theta_4 = y_j, (\theta_5 = c_5, \theta_6 = c_6, \theta_7 = c_7)\}$$

Where $[c_1$ & $c_2]$ and $[c_5, c_6$ & $c_7]$ are two non-identical group of constants to be evaluated using inverse kinematics solution for 'S' and 'G' respectively.
The general lemma in this regard is stated as below,

$$N_i \equiv \{(\theta_1 = c_1, \theta_2 = c_2, \ldots, \theta_{p-1} = c_{p-1}), \theta_p = x_i, \theta_q = y_i, (\theta_{q+1} = c_{q+1}, \theta_{q+2} = c_{q+2}, \ldots, \theta_k = c_k)\}$$

and

$$N_j \equiv \{(\theta_1 = c_1, \theta_2 = c_2, \ldots, \theta_{p-1} = c_{p-1}), \theta_p = x_j, \theta_q = y_j, (\theta_{q+1} = c_{q+1}, \theta_{q+2} = c_{q+2}, \ldots, \theta_k = c_k)\}$$

where, k: the degrees-of-freedom of the articulated robot; $\theta_1, \theta_2, \ldots \theta_p, \theta_q, \ldots, \theta_k$: joint-angles of the robot of which 'p' & 'q' represent any two consequtive pair; $[\theta_p -- \theta_q]$: the most significant c-space slice map; $[c_1, c_2, \ldots, c_{p-1}]$ & $[c_{q+1}, c_{q+2}, \ldots, c_k]$: two non-identical group of constants to be evaluated using inverse kinematics solution for 'S' and 'G' respectively.
Once this path is obtained for a particular slice, say the first one, the same procedure can be repeated for other slices too, because all the slices are identical. And, obviously the same path will be obtained in all slices, which will be designated as the final path for that robot in 3D. The kinematic inversion for 'S' & 'G' is another important facet in this regard and analytically, there can be multiple feasible positions of 'S' as well as 'G' in the c-space plot (refer fig. 25). One each from the two clusters of feasible locations can be selected for S: (α,β) and G: (γ,δ).

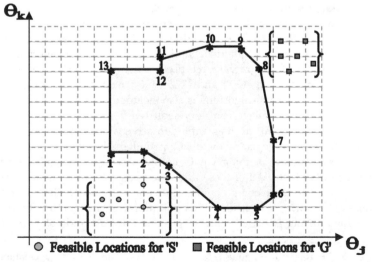

Fig. 25. Multiple feasible locations for 'S' & 'G' inside a specific c-space plot

4.5.2 Model for obtaining safe path for non-identical slices

In this case, we will get different sets of nodal points, corresponding to safe paths, for each slice. The procedure for obtaining a particular set of nodal points (nodes) for a specific slice is same as described in 4.5.1. But the challenge involved here is the most significant c-space slice is not fixed for the slices, rather in worst case it will differ. For example, let us take the case of seven d.o.f. manipulator and we assume there are five slices in the workspace. After c-space slice mapping, we find that while $[\theta_3 -- \theta_4]$ is the most significant map for the first slice, $[\theta_1 -- \theta_2]$ is the same for second slice. And likewise, the maps of $[\theta_3 -- \theta_4]$, $[\theta_5 -- \theta_6]$ & $[\theta_1 -- \theta_2]$ are the significant ones for third, fourth and fifth slice respectively. Thus the hurdle becomes in unifying these varying sets of maps into one final path. This is solved considering the *union* of the available sets of nodal points.

In general, if there are 'k' non-identical slices and the sets of nodal points for each safe path are represented by, $\{S_k\} = \{N_{1k}, N_{2k}, \ldots\ldots, N_{qk}\}$, where 'q' is the cardinality of the set and the value of 'q' may vary for different 'k', then the final path will be defined as,

$$\{S_{Final}\} = \bigcup_{p=1}^{p=k}\{S_p\}$$

In other words, statistical union of nodes will proceed slice-wise; i.e. to get the *safe* path complete posting all the nodes under one slice, then move on to the next slice and so on, till all the slices are exhausted. Nonetheless, the co-ordinates of the nodes in the path will be evaluated as per the lemma described in 4.5.1.

4.6 Illustrative examples

The developed path planning algorithm has been tested with two sample environments, the first one is contains a two degrees-of-freedom robot while the second one includes a seven degrees-of-freedom robot.

4.6.1 Sample workspace for two degrees-of-freedom robot

This example has a reference to the robot workspace with circular obstacles, as shown in fig. 6 and subsequently, the c-space map, vide fig. 8. Figure 26 presents the final c-space obstacles[7] with nodes numbered sequentially and the visibility graph generated there from.

Table 5 shows the output of the graph search process using our algorithm, developed with 'S' & 'G' configurations as $(60^0, -120^0)$ and $(222^0, 190^0)$ respectively. For comparison, the result obtained through A* search algorithm is also included.

It may be mentioned here that we can very well *benchmark* the result obtained by the Angular Deviation Algorithm, as it is representative amongst the *AI-based* heuristic algorithms. In fact, due to its logistics, the developed algorithm is having an edge over the other possible metrics of path planning, e.g. Generalized Voronoi Diagram (GVD), Cellular Automata and Potential Field. All of these three methods rely on diversification in search, which eventually leads to more computational time and complexity. Besides, the important attribute, namely, *"closeness to desired path"* is compromised in most of the non-AI based techniques. In comparison, AI-based searches are much robust and coherent; like the case

---

[7] Only obstacle 3,5 & 6, referred in figures 6 & 8, have been considered for the creation of visibility graph in order to reduce computational complexity.

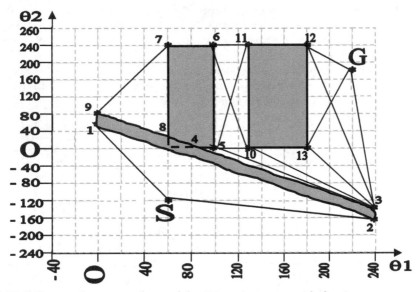

Fig. 26. Visibility graph generated out of the 2D environment, vide fig. 6

Algorithms Used	Path With Nodes	Joint-angle Combination (in degrees)
Angular Deviation Algorithm	S ⇒ 2 ⇒ 3 ⇒ G	(60,-120); (240,-166); (240,-139); (222,190)
A* Algorithm	S ⇒1⇒ 9 ⇒ 7 ⇒ 13⇒ 12 ⇒ G	(60,-120); (0,66); (0,86); (67,240); (129,240); (187,240); (222,190)

Table 5. Collision-free Near-optimal Path between 'S' & 'G' with reference to Example in fig. 26

with Angular Deviation Algorithm. The other group of search algorithms, based on mathematical programming, such as Variational Methods, Hierarchical Dynamic Programming and Tangent Graph Method, are although relatively better focused, but those are highly *memory-extensive*. Thus, in all counts, Angular Deviation Algorithm scores high amongst the various alternatives in graph-search methods.

4.6.2 Sample workspace with seven degrees-of-freedom robot

This example is in reference to the robotic environment shown in fig. 17 and subsequently the various c-space slice maps, as detailed in fig. 18. As we have declared in section 3.5 that $[\theta_1 -- \theta_2]$ plot is the most significant out of the four plots, we need to obtain the v-graph for this plot. Figure 27 shows the developed v-graph for this c-space slice plot. Here the generated v-graph is relatively simpler by default as it has only four nodes, which is due to the fact that both the joint-angles in consideration, viz. θ_1 & θ_2 are plotted in their full ranges. Thus 'S' & 'G' are also located on the boundary lines, because other locations will be infeasible.

However, it is to be noted that the exact shape of the v-graph (generated out of the most significant c-space slice plot) will depend upon the joint-angle ranges of the joint-pair under consideration and we will have distinct locations for 'S' & 'G', outside the c-space zone. It is evident from fig. 27 that S⇒2⇒3⇒G is the optimal path as 'α' is the smaller angle, which guides this path as per Angular deviation algorithm.

The generalized formulation for evaluating the angular position of 'S' & 'G' in the v-graph (using inverse kinematics routine for the manipulator) is as follows,

$$\sum_{j=1}^{j=7} l_j Cos\left(\sum_{k=1}^{k=j} \theta_k^m\right) = X_m \tag{43a}$$

$$\sum_{j=1}^{j=7} l_j Sin\left(\sum_{k=1}^{k=j} \theta_k^m\right) = Y_m \tag{43b}$$

where. 'm' : positional attribute of the end-point, i.e. either 'S' or 'G'; {l_j}: link-lengths; {θ_k^m}: joint-angles for 'S' or 'G' and (X_m, Y_m): planar Cartesian co-ordinates for 'S' or 'G'.

Now considering the Cartesian co-ordinates for 'S' as (20, 72.5) and the constant values for {$\theta_3\ \theta_4\ \theta_5\ \theta_6\ \theta_7$} as [$10^0\ 5^0\ 15^0\ 6^0\ 10^0$] we can solve for θ_1 & θ_2 using eqns. (43), which gives us $\theta_1^S \cong 60^0$ and $\theta_2^S \cong 0^0$. Similarly considering 'G' as (-30, -40) with the constant values for {$\theta_3\ \theta_4\ \theta_5\ \theta_6\ \theta_7$} as [$5^0\ 8^0\ 18^0\ 4^0\ 13^0$] we can solve for θ_1 & θ_2, which gives us $\theta_1^G \cong 108^0$ and $\theta_2^G \cong 120^0$.Thus, as proposed in section 4.5.1, the nodes, {$N_k,\ \forall k=1,2,3,4$} of the final collision-free path of the manipulator between 'S' & 'G' will be: $N_1 \equiv$ 'S' = (60^0, 0^0, 5^0, 8^0, 18^0, 4^0, 13^0); $N_2 =$ (140^0, 0^0, 5^0, 8^0, 18^0, 4^0, 13^0); $N_3 =$ (140^0, 120^0, 5^0, 8^0, 18^0, 4^0, 13^0) and $N_4 \equiv$ 'G' = (108^0, 120^0, 5^0, 8^0, 18^0, 4^0, 13^0).

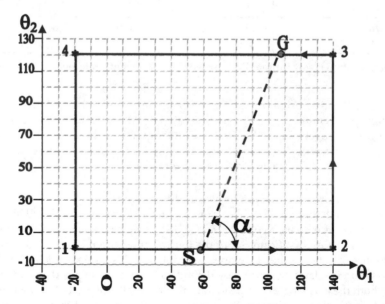

Fig. 27. Visibility graph obtained as per the most significant c-space slice plot of fig. 18

5. Case study

We have studied one real-life case of robot path planning in 3D, based on c-space modeling and v-graph searching, as delineated in the paper so far. The study was made with a five degrees-of-freedom articulated robot, RHINO-XR 3®, during its traverse between two pre-defined spatial locations through a collision-free path. The main focus was to maneuver this robot between 3D obstacles in reaching a goal location in a cluttered (laboratory) environment. Since RHINO is a low-payload robot, instead of standard pick-and-place tests, we designed our experiment such that it had to only *touch* the start ('S') and goal ('G') locations by the gripper end-point. The safe path in 3D, between the start and goal locations, was arrived using *c-space slice* mapping and *Angular Deviation Algorithm* (refer section 4.3). Figure 28 presents the photographic view of the experimental set-up, emphasizing the combined obstacle zone.

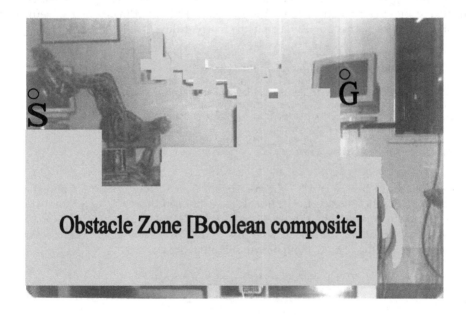

Fig. 28. Photographic view of the test set-up for spatial path planning with RHINO robot

Based on the obstacle zone map vis-à-vis waist rotation of the RHINO robot, we have discretized the workspace into three *non-identical* slices. The task-spaces, corresponding to these slices, are schematically shown in fig. 29. In all the sliced maps, the vertices of the combined obstacles are labeled alphabetically, with a numeric indication for the slice-number. For example, the vertex "A1" signifies the vertex number "A" in slice number1. It is to be noted that the vertex numbers are not *obstacle-specific*, rather those are serially numbered depending on the shape of the obstacle-zone in that very slice.

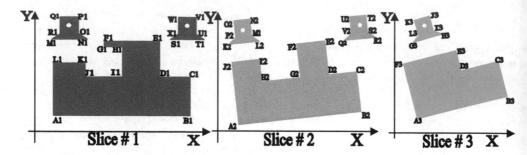

Fig. 29. Schematics of the Cartesian task-space slices for the case-study with RHINO robot

The co-ordinates of the vertices under each of the three slices were obtained by physical measurement of the task-space in 3D. In other words, first we took the measurements of the Boolean obstacle(s) in (x,y,z) form and then the planar co-ordinates of the slice-vertices were evaluated using the method of *projection geometry*. The co-ordinates of the vertices, so evaluated, under each slice, are tabulated in the matrix below.

*Slice*1 ⇒	{A1:(10,20)	B1:(190,20)	C1:(190,68)	D1:(145,68)	E1:(145,98)	F1:(85,98)	G1:(85,80)	H1:(95,80)
I1:(95,68)	J1:(45,68)	K1:(45,78)	L1:(10,78)	M1:(15,128)	N1:(45,128)	O1:(40,135)	P1:(40,155)	Q1:(20,155)
R1:(20,135)	S1:(210,128)	T1:(240,128)	U1:(235,135)	V1:(235,155)	W1:(215,155)	X1:(215,135)}		
*Slice*2 ⇒	{A2:(15,25)	B2:(195,35)	C2:(192,54)	D2:(170,58)	E2:(160,88)	F2:(142,68)	G2:(140,58)	I2:(45,68)
J2:(15,68)	K2:(18,130)	L2:(52,132)	M2:(47,138)	N2:(48,142)	O2:(22,150)	P2:(22,132)	Q2:(212,138)	R2:(242,140)
S2:(236,136)	T2:(236,156)	U2:(222,154)	V2:(220,134)}				
*Slice*3 ⇒	{A3:(18,28)	B3:(198,38)	C3:(193,56)	D3:(172,58)	E3:(168,88)	F3:(142,88)	G3:(18,134)	H3:(52,134)
I3:(46,138)	J3:(48,144)	K3:(24,148)	L3:(24,128)}				

The co-ordinates of the 'S' and 'G' are measured prior to the experiment and those are (30,145,40) & (225,145,42) respectively. It may be mentioned that 'S' & 'G' will not appear in the sliced task-space(s); rather, those will be only in 3D task-space as well as in c-spaces (sliced). Now, considering the kinematics of the RHINO robot, we get a feasible set of joint-angle combinations for 'S' & 'G' though inverse kinematics as,

$$\text{'S': } \{\theta_1 = 50^0, \theta_2 = -10^0, \theta_3 = 15^0, \theta_4 = 50^0, \theta_5 = 15^0\}$$

and

$$\text{'G': } \{\theta_1 = 60^0, \theta_2 = 120^0, \theta_3 = 58^0, \theta_4 = 98^0, \theta_5 = 20^0\}.$$

As evident by now, we will have three non-identical c-space slice maps for this environment and for the clarity in comparison between these three maps, we have selected common scale for the joint-angles. For example, the scale of 'θ_1' in *slice1* map will be same as that of in *slices 2 & 3* and likewise, for other joint-angles. This universality in scaling is helpful in judging the *most critical* map of a particular slice. We will now present the details of the three c-space slice maps, in their final form, alongwith the demarcation of the most critical map(s).

Figures 30,31 & 32 illustrate the conjugate maps corresponding to slice#1, slice#2 & slice#3 respectively.

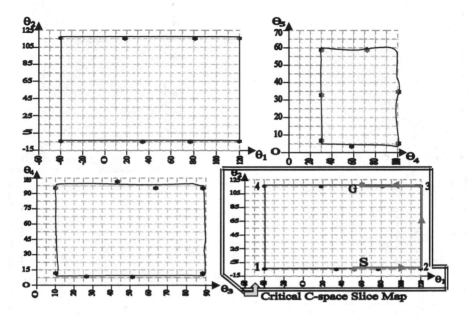

Fig. 30. C-space map for the first slice pertaining to the case- study

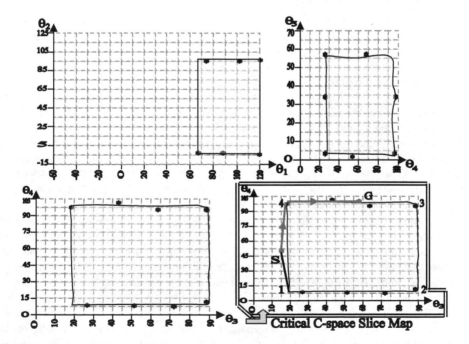

Fig. 31. C-space map for the second slice pertaining to the case – study

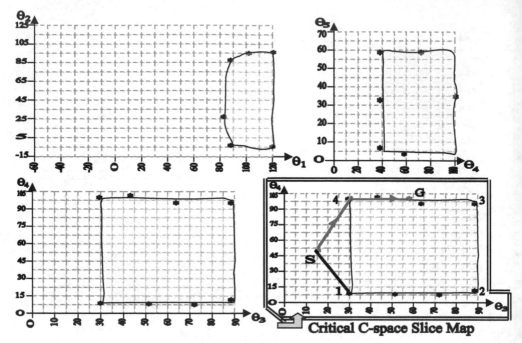

Fig. 32. C-space map for the third slice pertaining to the case – study

It is to be noted that some of the c-space slice maps in the above figures bear similarity; in fact, those maps are bounded by rectangular regions, occupying the full rotational ranges[8] of the participating joint-angles. That means, the full region is formidable, so far as the selection of safe nodes are concerned. This property is unique in the developed method, and it is helpful for obtaining the safe path in the final *go*. Now, assimilating all the three critical c-space slice maps as per figs. 30,31 & 32, we get the final safe path for the environment as the statistical *union* of the slice maps and it is represented as, $S \Rightarrow$ "2"[1] $\Rightarrow$ "3"[1] $\Rightarrow$ "4"[2] $\Rightarrow$ "4"[3]$\Rightarrow$G, where the legend "N"[k] symbolizes node 'N' of the k[th.] slice (refer section 4.5.2 for the formulation). Thus, the final path has got four *Intermediate Points* (IP), besides 'S' & 'G'. The joint-angle combinations for these 'IP$_x$', $\forall x=1,..,4$ (as labeled from 'S' onwards) are evaluated as, IP$_1\equiv$ "2"[1] $\Rightarrow\{\theta_1=120^0, \theta_2 =-10^0, \theta_3=58^0, \theta_4=98^0 \theta_5=20^0\}$; IP$_2 \equiv$"3"[1] $\Rightarrow\{\theta_1=120^0, \theta_2 =120^0, \theta_3=58^0, \theta_4=98^0 \theta_5=20^0\}$; IP$_3 \equiv$"4"[2] $\Rightarrow\{\theta_1=50^0, \theta_2 =-10^0, \theta_3=20^0, \theta_4=97^0 \theta_5=20^0\}$ and IP$_4$ $\equiv$"4"[3]$\Rightarrow \{\theta_1=50^0, \theta_2 =-10^0, \theta_3=30^0, \theta_4=97^0 \theta_5=20^0\}$.

Table 6 presents a summary of the various important outputs pertaining to the case study, with details of the computational time (for PC-based evaluation). Here, *Elapsed Time* has been divided into elemental time-periods (computational) against 4 sub-heads, viz. "**A**": *Generation of slices in task-space with co-ordinates (x,y) & node numbering;* "**B**": *Generation of c-space maps, including the critical-most;* "**C**": *Development of the v-graph* and "**D**": *Graph searching & output of the Angular Deviation Algorithm.*

---

[8] The *effective* rotational ranges of the five joint-angles of the RHINO robot are, θ1: (-40⁰ to 120⁰), θ2: (-10⁰ to 120⁰), θ3: (10⁰ to 90⁰), θ4: (10⁰ to 100⁰) and θ5: (5⁰ to 65⁰), as selected on the basis of our task-space layout & experiments.

Slice	Task Space	Critical C-space & V-graph path	Elapsed Time (for computation) [sec.]				
			A	B	C	D	Total
1		θ_1 vs. θ_2	4.5869	32.3878	16.8239	3.7116	57.5102
2		θ_3 vs. θ_4	4.3758	30.9764	16.7132	3.7208	55.7862
3		θ_3 vs. θ_4	3.8916	30.3358	16.4581	3.7211	54.4066
Final (Combined Computation) $\Rightarrow$			4.7938	32.8832	16.8423	3.7428	58.2621

Table 6. Summary Data for the Case Study with Details of the Computational Time [PC-based]

It may be noted that elemental timings for module A, B, C & D against individual slices give an apprehension regarding the relative toughness of the corresponding task-space and later on c-space & v-graphing. On the contrary, the final combined timings indicate the actual processing time (using multi-session processing of the operating system of the PC) of the problem, with usual co-processor actuations. Similar timings were observed while using A^* *algorithm* for graph search.

6. Conclusions

The details of the visibility graph-based heuristic algorithm for *safe* path planning in 2D plane as well as 3D space have been discussed in the paper, backed up by the theoretical paradigms of the generation of c-space obstacles from their respective task-spaces. The outcome of the c-space and v-graph algorithms have been found effective in programming the robot in order to perform certain pre-specified tasks or a series of tasks, such as in somewhat off-the-track industrial applications. The *best* path needs to be selected out of the possible alternatives by considering the most feasible criteria, which is essentially application specific. The novelty of the developed method lies with the ease of computational burden as 2D c-space slices are being joined statistically (*union*). Also by not incorporating all obstacles in one c-space slice we are improving upon computational efficiency and thereby reducing undue technical details regarding the obstacles. However, the safe path obtained by the developed method may overrule some nearer nodes, because the corresponding c-space slices are based on maximum *safety margins*, as per the propositions of the model. In fact, the concept of *formidable zones* is introduced in our model to avert potentially dangerous joint-angle configurations and thus, at times, the entire joint-angle range-space gets selected for the c-space map. The reason for taking this lemma is to safeguard the robot's motion between 'S' & 'G' to the best extent. Thus we may end up in some joint-angle (nodal) combinations, which might have been omitted, but it is always better to select a safe & secured path, rather than risking the robot motion for potential grazing and/or full collision with the obstacle(s). As per the proposed method, c-space slices often look trivial (e.g. regular geometrical shaped obstacles), although those are quite computationally intensive. Nonetheless, the geometrical simplification in appearance makes the v-graph map easy and subsequently the graph-search process too.

7. Acknowledgment

Gratitude is due to the faculties of the robotics laboratory, department of Production Engineering, Jadavpur University, Kolkata, India for supporting with the case study. The author acknowledges useful contribution made by Shri Sovan Biswas, Infosys Ltd., India in coding the path planning algorithm used in the case study.

8. References

Acar, Ercan U., Chosel, H., Zhang, Y. & Schervish, M., "Path Planning for Robotic Demining: Robust sensor-based Coverage of Unstructured Environments and Probabilistic Methods", The International Journal of Robotics Research, vol. 22, no. 7-8, July-August 2003, pp 441-466.

Bajaj, C. & Kim, M.S., "Generation of Configuration Space Obstacles: Moving Algebraic Surfaces", The International Journal of Robotics Research, vol. 9, no. 1, February 1990, pp 92-112.

Branicky, M.S. & Newman, W.S., "Rapid Computation of Configuration Space Obstacles", Proceedings of the IEEE International Conference on Robotics & Automation, 1990, pp 304-310.

Brooks, R.A., "Solving the Find-path Problem by Good Representation of Free Space", IEEE Transactions on Systems, Man & Cybernatics, vol. SMC-13, no. 3, 1983, pp 190 - 197.

Brost, R.C., "Computing Metric and Topological Properties of Configuration Space Obstacles", Proceedings: IEEE International Conference on Robotics & Automation, 1989, pp 170-176.

Campbell, D. & Higgins, J., "Minimal Visibility Graphs", Information Processing Letters, vol. 37, no. 1, 10th. January 1991, pp 49-53.

Curto, B. & Moreno, V., "Mathematical Formalism for the Fast Evaluation of the Configuration Space", Proceedings of the IEEE International Symposium on Computational Intelligence in Robotics and Automation, Monterey, CA, U.S.A., July 1997, pp 194-199.

De Pedro, M.T. & Rosa, R.G., "Robot Path Planning in the Configuration Space with Automatic Obstacle Transformation", Cybernatics & Systems, vol. 23, no. 4, 1992, pp 367 - 378.

Erdmann, Michael, "On a Representation of Friction in Configuration Space", The International Journal of Robotics Research, vol. 13, no. 3, June 1994, pp 240-271.

Fu, Li-Chen & Liu, Dong-Yuch, "An Efficient Algorithm for Finding a Collision-free Path Among Polyhedral Obstacles", Journal of Robotics Systems, vol. 7, no.1, 1990, pp 129-137.

Gilbert, E.G. & Johnson, D.W., "Distance Functions and Their Application to Robot Path Planning in the Presence of Obstacles", IEEE Transactions on Robotics & Automation, vol. RA-1, no. 1, March 1985, pp 21-30.

Hasegawa, T. & Terasaki, H., "Collision Avoidance: Divide - and - Conquer Approach by Space Characterization and Intermediate Goals", IEEE Transactions on Systems, Man & Cybernatics, vol. SMC-18, no. 3, May-June 1988, pp 337 - 347.

Hwang, Y.K. & Ahuja, N., "Gross Motion Planning - A Survey", ACM Computing Surveys, vol. 24, no. 3, 1992, pp 219-291.

Jun, S. & Shin, K.G., "A Probabilistic Approach to Collision-free Robot Path Planning", Proceedings of the IEEE International Conference on Robotics & Automation, 1988, pp 220-225.

Keerthi, S.S. & Selvaraj, J., "A Fast Method of Collision Avoidance For An Articulated Two Link Planar Robot Using Distance Functions", Journal of the Institution of Electronics & Telecommunication Engineers, vol. 35, no. 4, 1989, pp 207-217.

Khouri, J. & Stelson, K.A., "Efficient Algorithm for Shortest Path in 3-D with Polyhedral Obstacles", Transactions of the ASME - Journal of Dynamic Systems, Measurement & Control, vol. 8, no. 3, September 1989, pp 433-436.

Kohler, M. & Spreng, M., "Fast Computation of the C-space of Convex 2D Algebraic Objects", The International Journal of Robotics Research, vol. 14, no. 6, December 1995, pp 590-608

Lozano-Perez', T., "Spatial Planning: A Configuration Space Approach", IEEE Transactions on Computers, vol. C-32, no. 2, 1983, pp 108-120.

Tomas Lozano-Perez', "A Simple Motion Planning Algorithm for General Robot Manipulators", IEEE Transactions on Robotics & Automation, vol. RA-3, no. 3, June 1987, pp 207-223.

Lumelsky, V. & Sun, K.,"A Study of the Obstacle Avoidance Problem Based on the Deformation Retract Technique", Proceedings of the 29th. IEEE Conference on Decision and Control, Honolulu, HI, U.S.A., Dec. 1990, pp 1099-1104.

Lumelsky, V. & Sun, K, "A Unified Methodology for Motion Planning with Uncertainty for 2-D and 3-D Two-link Robot Arm Manipulators", The International Journal of Robotics Research, vol. 9, no. 5, October 1990, pp 89-104.

Ralli, E. & Hirzinger, G., "Global and Resolution Complete Path Planner for up to 6 dof Robot Manipulators", Proceedings of the IEEE International Conference on Robotics & Automation, Minneapolis, MN, U.S.A., April 1996, pp 3295-3302.

Red, R.E. & Truong-Cao, H.V, "Configuration Maps for Robot Path Planning in Two Dimensions", Transactions of the ASME - Journal of Dynamic Systems, Measurement & Control, vol. 107, December 1985, pp 292-298.

Red, W.E. et al, "Robot Path Planning in Three Dimensions Using the Direct Subspace", Transactions of the ASME - Journal of Dynamic Systems, Measurement & Control, vol. 119, September 1987, pp 238-244.

Roy, D., "Study on the Configuration Space Based Algorithmic Path Planning of Industrial Robots in an Unstructured Congested 3-Dimensional Space: An Approach Using

Visibility Map", Journal of Intelligent and Robotic Systems, vol. 43, no. 2– 4, August 2005, pp 111-145.

Sacks, E. & Bajaj, C., "Sliced Configuration Spaces for Curved Planar Bodies", The International Journal of Robotics Research, vol. 17, no. 6, June 1998, pp 639-651.

Sacks, E., "Practical Sliced Configuration Spaces for Curved Planar Pairs", The International Journal of Robotics Research, vol. 18, no. 1, January 1999, pp 59-63.

Sachs, S., La Valle, S.M. & Rajko, S., "Visibility-based Pursuit – Evasion in an Unknown Planar Environment", The International Journal of Robotics Research, vol. 23, no. 1, January 2004, pp 3-26.

Schwartz, J.T. & Sharir, M., "A Survey of Motion Planning and Related Geometric Algorithms", Artificial Intelligence, vol. 37, 1988, pp 157-169.

Slotine, Jean Jacques, E. & Yang, H.S., "Improving the Efficiency of Time-optimal Path Following Algorithm", IEEE Transactions on Robotics & Automation, vol. RA-5, no. 1, 1989, pp 118-124.

Verwer, Ben J.H., "A Multi-resolution Workspace, Multi-resolution Configuration Space Approach to Solve the Path Planning Problem", Proceedings of the IEEE International Conference on Robotics & Automation, 1990, pp 2107-2112.

Welzl, E., "Constructing the Visibility Graph for n-line segments in $O(n^2)$ Time", Information Processing Letters, vol. 20, Sept. 1985, pp 167-171.

Wise, Kevin D. & Bowyer, A., "A Survey of Global Configuration-space Mapping Techniques for a Single Robot in a Static Environment", The International Journal of Robotics Research, vol. 19, no. 8, August 2000, pp 762-779.

Yu, Y. & Gupta, K., "C-space Entropy: A Measure for View Planning and Exploration for General Robot - - Sensor Systems in Unknown Environment", The International Journal of Robotics Research, vol. 23, no. 12, December 2004, pp 1197-1223.

Zelinsky, A., "Using Path Transforms to Guide the Search for Findpath in 2D", The International Journal of Robotics Research, vol. 13, no. 4, August 1994, pp 315-325.

Permissions

The contributors of this book come from diverse backgrounds, making this book a truly international effort. This book will bring forth new frontiers with its revolutionizing research information and detailed analysis of the nascent developments around the world.

We would like to thank Assoc. Prof. PhD. Serdar Küçük, for lending his expertise to make the book truly unique. He has played a crucial role in the development of this book. Without his invaluable contribution this book wouldn't have been possible. He has made vital efforts to compile up to date information on the varied aspects of this subject to make this book a valuable addition to the collection of many professionals and students.

This book was conceptualized with the vision of imparting up-to-date information and advanced data in this field. To ensure the same, a matchless editorial board was set up. Every individual on the board went through rigorous rounds of assessment to prove their worth. After which they invested a large part of their time researching and compiling the most relevant data for our readers. Conferences and sessions were held from time to time between the editorial board and the contributing authors to present the data in the most comprehensible form. The editorial team has worked tirelessly to provide valuable and valid information to help people across the globe.

Every chapter published in this book has been scrutinized by our experts. Their significance has been extensively debated. The topics covered herein carry significant findings which will fuel the growth of the discipline. They may even be implemented as practical applications or may be referred to as a beginning point for another development. Chapters in this book were first published by InTech; hereby published with permission under the Creative Commons Attribution License or equivalent.

The editorial board has been involved in producing this book since its inception. They have spent rigorous hours researching and exploring the diverse topics which have resulted in the successful publishing of this book. They have passed on their knowledge of decades through this book. To expedite this challenging task, the publisher supported the team at every step. A small team of assistant editors was also appointed to further simplify the editing procedure and attain best results for the readers.

Our editorial team has been hand-picked from every corner of the world. Their multi-ethnicity adds dynamic inputs to the discussions which result in innovative outcomes. These outcomes are then further discussed with the researchers and contributors who give their valuable feedback and opinion regarding the same. The feedback is then collaborated with the researches and they are edited in a comprehensive manner to aid the understanding of the subject.

Apart from the editorial board, the designing team has also invested a significant amount of their time in understanding the subject and creating the most relevant covers. They scrutinized every image to scout for the most suitable representation of the subject and create an appropriate cover for the book.

The publishing team has been involved in this book since its early stages. They were actively engaged in every process, be it collecting the data, connecting with the contributors or procuring relevant information. The team has been an ardent support to the editorial, designing and production team. Their endless efforts to recruit the best for this project, has resulted in the accomplishment of this book. They are a veteran in the field of academics and their pool of knowledge is as vast as their experience in printing. Their expertise and guidance has proved useful at every step. Their uncompromising quality standards have made this book an exceptional effort. Their encouragement from time to time has been an inspiration for everyone.

The publisher and the editorial board hope that this book will prove to be a valuable piece of knowledge for researchers, students, practitioners and scholars across the globe.

List of Contributors

Weiwei Shang and Shuang Cong
University of Science and Technology of China, P.R. China

Metin Aydin
Kocaeli University, Department of Mechatronics Engineering, Kocaeli, Turkey

Przemysław Mazurek
Department of Signal Processing and Multimedia Engineering, West Pomeranian University of Technology, Szczecin, Poland

Anna Walaszek-Babiszewska
Opole University of Technology, Poland

Pakize Erdogmus
Duzce University, Engineering Faculty, Computer Engineering Department, Turkey

Metin Toz
Duzce University, Technical Education Faculty, Computer Education Department, Duzce, Turkey

Özer Çiftçioğlu and Sevil Sariyildiz
Delft University of Technology, Faculty of Architecture, Delft, the Netherlands

Rogério Rodrigues dos Santos and Valder Steffen Jr.
School of Mechanical Engineering, Federal University of Uberlândia, Uberlândia, MG, Brazil

Sezimária de Fátima Pereira Saramago
Faculty of Mathematics, Federal University of Uberlândia, Uberlândia, MG, Brazil

Nguyen Minh Thanh
Department of Automation, Hochiminh City University of Transport, Vietnam

Le Hoai Quoc
Department of Science and Technology, People's Committee of Hochiminh City, Vietnam

Victor Glazunov
Mechanical Engineering Research Institute, Russian Academy of Sciences, Russia

M. Falahian, H.M. Daniali and S.M. Varedi
Babol University of Technology, Iran

Debanik Roy
Board of Research in Nuclear Sciences, Department of Atomic Energy, Government of India, Mumbai, India

Printed in the USA
CPSIA information can be obtained
at www.ICGtesting.com
JSHW011423221024
72173JS00004B/658

9 781632 384027